U0898612

环筑

ENVIRONMENT&ARCHITECTURE

仙田 满＋环境设计研究所 著
Text: Mitsuru Senda+Environment Design Institute

水晶石数字传媒 编
Edit: Crystal Digital Media Company

中国建筑工业出版社
China Architecture & Building Press

图书在版编目（CIP）数据

环筑：仙田 满＋环境设计研究所著 / 水晶石数字传媒编.
－北京：中国建筑工业出版社，2003
ISBN 7-112-06086-9

Ⅰ.环…Ⅱ.水…Ⅲ.建筑设计－作品集－日本－现代 Ⅳ.TU206

中国版本图书馆 CIP 数据核字(2003)第 093957 号

- 摄　影 Photo　藤冢光政（不包含标有出处的图片） Mitsumasa Fujitsuka (Except the Photos with Notice)
- 翻　译 Translator　王 昀 Wang Yun　宁 晶 Ning Jing
  巴贝尔翻译传媒中心 Babel Trans-Media Center　环境设计研究所 Environment Design Institute
- 责任编辑 Editor　黄居正 Huang Juzheng　李迪悃 Li Dikun
- 特约编辑 Contributing Editor　李晓鸿 Li Xiaohong　杨熹微 Yang Xiwei
- 封面设计 Cover　徐 黎 Xu Li
- 装帧设计 Layout　水晶石数字传媒 Crystal Digital Media Company

环　筑
ENVIRONMENT&ARCHITECTURE

仙田 满＋环境设计研究所　著
Text: Mitsuru Senda+Environment Design Institute

水晶石数字传媒　编
Edit: Crystal Digital Media Company

*

中国建筑工业出版社出版、发行（北京西郊百万庄）
新华书店经销
精美彩色印刷有限公司印刷

*

开本：787 × 1092 毫米 1/12 印张：26
2003 年 11 月第一版 2003 年 11 月第一次印刷
印数：1 － 3000 册　定价：298.00 元
ISBN 7-112-06086-9
TU · 5352(12099)

(邮政编码：100037)
本社网址: http://www.china-abp.com.cn
网上书店: http://www.china-building.com.cn

目录 | 目次 | General Contents

序

马国馨　中国工程院院士
北京市建筑设计研究院顾问总建筑师

既是建筑师又是环境建筑师的日本朋友仙田满先生将在中国出版他的专集：《环筑——仙田满＋环境设计研究所》，通过本书的出版，我们可以直接了解仙田满先生35年来在建筑设计和环境设计领域的成就和经验，对于努力和国外建筑界交流的中国建筑师来说，肯定会有所启发和帮助。

对仙田先生开始有所印象还是缘于一件十分偶然的事情：每当我们去日本访问，从成田机场去东京的途中，都可以看到一栋花瓣形状的游泳馆，我还曾在飞驰的车辆中拍过照片。后来才知道这就是由仙田先生设计的辰巳国际游泳场。近年来有幸与仙田先生会面，那时他以日本建筑学会会长的身份访京，当时学会的几位领导不在北京，于是我在北京市建筑设计研究院接待了仙田先生和夫人，对他的创作有了进一步的了解，并承他赠送专著《环境设计的展开——概念和过程》，此后我又陆续整理翻阅了我手中的相关材料，这样开始对仙田先生的创作和理论有了进一步的领会。

仙田先生1941年出生于神奈川县。在日本建筑界像他这样战争中出生，成长于战后，开始从业于东京奥运和经济腾飞，并经历了泡沫经济破灭，经济的艰难复苏的同一年龄段的建筑师，有一大批是当今日本建筑界的重要人物，如叶祥荣（1940年）、安藤忠雄（1941年）、伊东丰雄（1941年）、早川邦彦（1941年）、长谷川逸子（1941年）、六角鬼丈（1941年）、石山修武（1944年）、石井和纮(1944年)、山本理显（1945年）……他们现在都精力充沛，经验丰富，活跃在国内外设计第一线。另外，日本建筑界也很讲究出身和师承。仙田先生1964年毕业于东京工业大学建筑系，当时著名的建筑师谷口吉郎（1904年－1979年）就在东工大任教。仙田在大学中曾致力于通过对城和茶室的研究来看历史空间的影响。毕业后即进入菊竹清训事务所工作过四年。菊竹先生在日本现代建筑史中是具有一定地位的建筑师，除了参与20世纪60年代的“新陈代谢运动”外，当时设计的出云大社厅舍（1963年）和东光园（1965年），也在建筑界引起很大反响。所以仙田先生在一次调查中把菊竹先生和他设计的出云大社厅舍列为自己最尊敬的建筑师和最喜爱的作品之一。前面提到的伊东丰雄和长谷川逸子也先后在菊竹的事务所工作过。此后仙田先生除在环境设计研究所的工作外，还曾在琉球大学、名古屋工业大学、东京工业大学任教，并获得过国际和国内的一系列奖赏。

仙田先生对于环境设计的研究和创见也引起了我的注意，因为这也是我在设计工作中比较关心，并经常探讨的一个课题。记得1986年中国的建筑界和美术界开展过一次关于环境设计和环境艺术的讨论，讨论十分热烈。我也曾在报刊上发表过几篇文章，其主要观点是：强调环境设计。“环境设计可称作是包罗万象的综合艺术，人们不是单纯地研究一栋房子孤立的形态美，而是更多地注重一个群体，许多群体结构和一个整体。”我认为“环境设计”比当时“环境艺术”、“景观设计”的提法更广泛，更全面，包容力更强。“人们更加认识到离开了环境，建筑本身就没有意义。建筑势必受环境的制约，受其他环境因素的影响。”“环境设计是时间和空间的结合……一个成熟、美观的环境的形成至少需要几十年甚至上百年的努力。”但由于各种原因，对这一课题的研究并未进一步深入。

而仙田先生在这一领域的研究就很有启发性和开创性，他的许多观点是积数十年实践的经验和体会所总结出的真知，下面我想列举一些主要观点（只是大意，并非原文）。

环境设计是关系设计。此关系就是空间的关系、时间的关系、社会的关系，是将空间、时间和社会关系进行再构筑。

环境设计是贯通设计。这一设计的领域大到地球设计，之后涉及国土设计、城市、地域、土木、造园、建筑、室内、陈设、产品、工业设计、图形、照明和装置设计等从大空间到小平面等各个领域。在逐渐专门化、专业化的设计界中必须注意对此加以考虑和运筹。

环境设计是综合设计。必须注重从系统调查、研究、企划、规划、设计、监理到运营一条龙的运作方式，特别是研究和设计的一体化十分重要。

环境设计是参与设计、协同设计和社会设计。要求业主、使用方及优秀的设计师一起协同工作，互相沟通，激发彼此的创造力，从而寻求环境的最佳解决方式。

环境设计是反馈设计。必须通过事先和事后的调查来使设计日趋完善。

从仙田先生众多的设计和策划实例中也可以发现如前所述的大处着眼、小处着手、统筹全局、互相协同的设计概念的过程。仅“儿童的游戏环境”这一课题就持续进行了三十年。日本同行在每一设计项目中（不管规模大小）精心推敲、一丝不苟的敬业和执着精神值得我们学习。同时在当前我国各地陆续修建科技馆、少年宫和文化中心的热潮中，仙田先生在几个科学馆、自然博物馆、儿童未来馆等公共建筑中的构思和经验也可供我们借鉴。

仙田先生除建筑师外还自称“环境建筑师”，他自称喜欢把自己的许多概念加以命名，像本书“环筑”一词即源于他自己创作的“环境建筑”（ENVIRONMENT AND ARCHITECTURE）的概念，仙田先生希望终有一日“环筑”也能和环境设计一样成为一个十分普通的词汇。我们希望，创造更好的环境和美好的环境建筑能成为建筑师们的共识。

推薦の言葉

馬国馨　院士　中国工程院
顧問総建築師　北京市建築設計研究院

建築家であり環境建築家である日本の友人の仙田満先生は、中国で作品集「環築—仙田満＋環境デザイン研究所」を出版されることとなった。本書の出版を通して、我々が直接先生の35年に及ぶ建築・環境デザインの領域でのご活躍と経験を知ることができ、海外の建築家との交流を盛んに重ねている中国の建築家にとって、大きな励みとなるにちがいない。

先生との最初の出会いは、偶然のきっかけであった。日本を訪問する度、成田空港から東京へ向かう途中花びらのような水泳場を目にしていて、激走する車の中から写真を撮るほどであった。後にそれが、先生が設計した東京辰巳国際水泳場だと知ったのだった。先生にお目にかかることができたのは、日本建築学会会長として北京に来訪されたときのことだった。学会責任者があいにく不在中で、私が北京建築設計研究院で仙田先生ご夫妻を接待した。先生の作品について一層知ることができ、著書の「環境デザインの展開」を贈られた。後日関連資料を手にし、仙田先生の作品および理論をより深く理解することができた。

仙田先生は1941年神奈川県で生まれた。日本の建築界において、彼のように戦争の時代に生まれ、戦後の成長期を経て、東京オリンピックと経済高度成長期で仕事に就き、経済バブル崩壊の時期をともに歩んできている同年代の建築家はほかにも多く現れている。たとえば葉祥栄（1940）、安藤忠雄（1941）、伊東豊雄（1941）、早川邦彦（1941）、長谷川逸子（1941）、六角鬼丈（1941）、石山修武（1944）、石井和紘（1944）、山本理顕（1945）等々。彼らは現在でも非常に元気で、日本国内外で設計の第一線で活躍している。また日本建築界において、その出身は非常に重要である。仙田先生は1964年東京工業大学を卒業されているが、著名な建築家である谷口吉郎（1904－1979）が当時東京工業大学で教えていた。仙田先生は在学中、城と茶室の研究を通して歴史が空間に及ぼす影響を研究した。卒業後は菊竹清訓氏の事務所で4年の実務を積んでいる。菊竹先生は日本の現代建築史において重要な地位を持つ建築家であり、60年代の「新陳代謝運動」の参加以外にも、出雲大社庁舎（1963）と東光園（1965）を設計し、大きな反響を呼んでいた。仙田先生はある調査の中で、菊竹先生と出雲大社庁舎を自分のもっとも尊敬している建築家と好きな建築として挙げている。上記で挙げた伊東豊雄氏および長谷川逸子氏も菊竹事務所で働いている。後に仙田先生は環境デザイン研究所を立ち上げ、琉球大学、名古屋工業大学、東京工業大学などで教育・研究活動に従事され、国内および海外でも作品が評価されて一連の賞を受賞されている。

仙田先生の環境を配慮した設計の研究やアイデアなども私の注意を引いた。これは私が設計する上で、常に関心を持ち討議する課題だからである。1986年に中国の建築業界と美術業界では、「環境設計」または「環境芸術」に関して熱烈な議論が行われた。私も当時このテーマについて何篇もの論文を発表し、環境設計の重要性を強調した。「環境設計はあらゆるものを含んでいる。人々は単純に1棟の建物の外観の美しさを追求するものではなく、「集合」、「集合のまとまり」、「全体」により多くの注意を払っている」などを主張した。「環境設計」が包括する範疇は、当時提唱された「環境芸術」及び「景観設計」に比べ広く、全般的に考慮でき、包容力も強いと考える。「周辺の環境を考慮しないと、建物そのものの意義がない。建物は環境に制限されると共に、他の環境要素に影響を及ぼされる。」また、「環境設計は時間と空間を結合した産物であり…、完備かつ美しい環境を形成するのに少なくとも数十年、百年以上の努力を続けなければならない。」などを考えていた。しかし様々な要因からこの課題に関する研究は続けていなかった。

仙田先生の研究は、この領域の中で我々の考えを啓発するものであり、先生の環境デザインに対する多くの考え方は、数十年の経験と体験を積重ねてまとめられた正しい認識である。以下にその主要な見解を挙げる。（原文ではなく概要である。）

・環境デザインは様々な関係を設計するものであり、空間の関係、時間の関係、社会の関係を含み、さらに、空間、時間及び社会関係を構築する。

・環境デザインはあらゆるものを結び付ける設計である。大きく言えば、地球を設計する。細かく考えると、国土都市、地域、土木、建築、室内、展示、プロダクト、工業デザイン、サイン、照明、機械設計など大空間から小空間までそれぞれの領域に及ぼす。専門化・分業化されている設計業界では、それらの関連性と動きに注意を払わなければならない。

・環境デザインは総合的な設計であるため、調査、研究、計画、設計、設計監理及び運営に至るまでの一連の仕事を系統化しなければならない。特に研究と設計を一体化することは大変重要である。

・環境デザインは参与する設計、協同する設計及び社会設計である。施主、利用者ならびに素晴らしい設計家が、共に働き交流し創造力を刺激しあいながら、よりよい環境づくりのための方策を探るものである。
・環境デザインは反映設計であり、設計前と設計後の調査により設計を検証し、最善のデザインを目指すべきである。
仙田先生が設計及び計画されたた多くの事例から、前述のような大局に着眼し、細微に工夫を凝らし、全体性を考慮し、協同して作り上げる設計という概念が明らかになってくるであろう。「こどもの遊び環境」という1つの課題で30年間続けてこられたことからでも、仙田先生が全ての設計（規模の大きさに関わらず）において、真摯な態度で弛むことなく臨んでいる姿勢に感銘を受けている。現在、わが国では科学館、青少年センター、市民文化センターを多く建造している。仙田先生は幾つかの科学館、自然博物館、児童未来館などの公共建築を設計されておられるが、その発想と経験は我々設計者の参考になると思われる。
最後に、仙田先生は建築家の他に、「環境建築家」と自称されている。先生は自分の提案された概念に名前を付けることが好きで、例えば本書の「環築」も先生が創作された「環境建築」（Environment and Architecture）という概念から略称したものである。近い将来では、「環築」という言葉も「環境設計」のようによく使われるものになることを望んでいる。より快適な環境及び美しい環境建築を造ってゆくことを、すべての建築家に認識させることは我々の望みである。

Recommendation

Ma Guoxin Academician Chinese Academy of Engineering
Senior Chief Architect Beijing Institute of Architect Design & Research

Architect Mitsuru Senda, who is also an environmental architect and friend of mine, is going to publish his new book titled 'Environment and Architecture - Mitsuru Senda + Environment Design Institute' in China. This book will certainly help Chinese architects who try to change their experience with their foreign counterparts know more about Mr. Mitsuru Senda's 35 years of achievements in architectural and environmental designs.

It was coincidental that Mr. Mitsuru Senda and I came to know each other. Whenever we went Japan, we could see a petal-shape swimming pool on the way from Narita Airport to Tokyo. I even took pictures of it in the car. Only later was I told it was the Tokyo Tatsumi International Swimming Center designed by Mitsuru Senda. Later on, Mr. Mitsuru Senda visited Beijing in the capacity of the president of Japanese Institute of Architecture. Since the officials of the Chinese Architectural Society were not in Beijing at the time, I received Mr. and Mrs. Senda at the Beijing Architectural Design Institute. During the meeting, he gave me his book 'Unfolding of Environmental Design - Concept and Process' and, after reading some information about him later on, I gradually knew more about his architectural theories.

Born in Yokohama City in 1941, Mr. Mitsuru Senda grew up during the war time and, when the war was over, started his works after the time of Tokyo Olympic Games and during rapid economic development. He also experienced the debacle of the bubble economy and difficult economic recovery. Many influential architects in Japan have the same experience as his such as Shoei Yoh (1940), Tadao Ando (1941), Toyo Ito (1941), Kunihiko Hayakawa (1941), Itsuko Hasegawa (1941), Kijyo Rokkaku (1941), Osamu Ishiyama (1944), Kazuhiro Ishii (1944), Riken Yamamoto (1945) and so on. They are experienced, full of vitality and very actively involved in the architectural circles both in Japan and abroad. Japanese architects attach importance to the master-disciple tradition. Mr. Mitsuru Senda graduated from the Department of Architecture, Tokyo Institute of Technology in 1964, where famous architect Yoshiro Taniguchi (1904-1979) was a professor. Mitsuru Senda, in his college years, tried to understand the influence of historical space through research on Japanese castles and teahouses. After graduation, he worked for 4 years at Kiyonori Kikutake's Architect Firm. Mr. Kikutake is an important architect in the modern architectural history of Japan, not only taking part in the 'Metabolism Movement' of the 1960s, but also designing the Izumo Taisha Hall (1963) and Toko Garden (1965) which made quite a stir at the time. Therefore, in one of his investigation, Mr. Mitsuru Senda regarded Mr. Kikutake and his Izumo Taisha Hall as his most admired architect and architectural work respectively. The above-mentioned Toyo Ito and Itsuko Hasegawa also once worked at Kikutake's firm. Later on, apart from establishing the Environment Design Institute, Mitsuru Senda also taught at Ryukyu University, Nagoya Institute of Technology and Tokyo Institute of Technology, and has won many domestic and international awards.

I was especially attracted by Mr. Mitsuru Senda's environmental design research and innovation, for I am concerned with and exploring this topic too. I recall that in 1986, there was a heated debate on environmental design and art among Chinese architects and artists. I published several articles expressing my view to espouse environmental design. I said, "environmental design is an all-embracing art by which the beauty of a house is not studied in an isolated way. Instead, focus is on groups, on many group structures and treat them as a whole. I believe the term 'environmental design' is more comprehensive and effective than 'environmental art' or 'landscape design'. People have realized that architecture would lose its meaning without environment, so architecture must be subject to environment and its impact. Environmental design is the combination of time and space. The formation of a mature and beautiful environment takes at least dozens of or even over a hundred years' effort." But out of various reasons, the research on this line was not further pursued.

However, Mr. Mitsuru Senda's research on this respect is very insightful because many of his views were based on his years of practice and experience. I would like to illustrate some of them in the following (only roughly, not quoted from his original words):

· Environmental design is deign of relations. Relation of space, of time, of society. It is to reconstruct space, time and social relations.

· Environmental design is a linking-up design which could range from earth design, in details to land design, city design, regional design, landscape, architecture, interior, display, products, industrial design, signs, lighting and installation. It embraces both large and small plane areas. In the more specified designing circle, emphases should be laid on the trend of comparison and summary.

· Environmental design is a comprehensive design which should coordinate systematic investigation, research, planning, design, designing supervision, especially the integration of research and design.

·Environmental design is design of involvement, coordination and social design. It requires the coordination of proprietors, users and good architects. The creativity of these groups of people should be stimulated so as to seek the best solutions to environmental issues.

· Environmental design also means feedback. Designs must be perfected by prior and post investigations.

Many of Mitsuru Senda's designs and planning demonstrate the concept of focusing on main issues, handling details scrupulously, coordination and taking the whole into consideration. The design and research project of 'children's play environment' alone took him as long as over 30 years and still continue. We should learn from our Japanese counterpart the professional dedication and meticulous care shown in each of their project regardless of its scale. As we are in a rush in building science museum, children's centers and cultural centers, we can draw upon some of Mitsuru Senda's concepts pertaining these buildings as well as science museums, nature museum, children's centers and museums.

Apart from architect, Mr. Mitsuru Senda also likes to call himself an environmental architect. He is fond of naming his concepts, for example, the term 'Environment and Architecture' for this book. He hopes this term will become a common word one day as that of environmental design. We hope that architects can join hands towards fulfilling the goal of creating a better environment and beautiful environmental architecture.

仙田满的建筑

伊东丰雄　　　建筑师

仙田满的建筑是幻想曲的表现。巨大的植物像花一样伸展和奔放。为什么会产生这样的幻想是有理由的。

我是大学毕业之后在菊竹清训的办公室里见到仙田的。当时他设计了名为「孩子的王国」的主题公园。小花瓣一样形状的小屋罗列在大自然中，在我的印象中那是极富诗意的设计。

也正是以此为契机，他在独立开业之后设计了许多为儿童准备的游戏道具。正是这种设计体验，孕育了建筑一般性所不能制约的奔放的表现力。

在中国这一广阔的土地上实现他的建筑的时候，他的这种幻想曲肯定将感动更多的人。

仙田満の建築

伊東豊雄　　　建築家

仙田満の建築はいつもファンタジーに溢れた表現を示している。巨大な植物が花開いたように伸びやかで奔放である。何故このようなファンタジーを持ちうるのか。それには理由がある。

私は大学を卒業した直後に働いた菊竹清訓のオフィスで仙田に出会った。その頃彼は「こどもの国」というテーマパークの施設をデザインしていた。小さな花弁のような形をしたロッジが自然のなかにいくつも並ぶ、きわめて詩的なデザインであったことを記憶している。それが契機となって、彼自身のアトリエを始めてからも、彼は子供のための遊戯施設や遊具を数多くデザインしていた。そうした経験が、建築の慣習に規制されない奔放な表現力を育んだように思われる。

中国という広大な土地において彼の建築が実現する時、そのファンタジーはより拡張されて人々を魅惑してくれるに違いない。

Architecture of Mitsuru Senda

Toyo Ito　　　Architect

Architecture of Mitsuru Senda always shows expression with full of fantasy. It is free and unrestrained as if a giant flower blooms. Why can his design have such a fantasy? There is a reason.

I met him at the office of Kiyonori Kikutake where I worked for just after I had graduated the university. In those days, he was designing a theme park, called “Children’s Land”. I remember that he proposed very poetic design which several small-petal shaped lodges stood in nature. Since then, after he started his own atelier, he designed many children’s facilities and play structures. Those experiences must have grown his free and unrestrained expression which was never regulated by the custom of architecture.

When his architecture is realized in China which has vast lands, the fantasy would be extended and must attract people.

植田实　　　　　　主编　住居的图书馆出版局

仙田满担任东京工业大学建筑系教授近十年来，致力于多项研究和建筑设计的同时，至今春末还担任了日本建筑学会会长职务，掌管着日本建筑界全体的事务，所以可称为日本现代建筑的中坚人物。但以我个人的观点来看，真正的他是远离那些职务，有着完全个人自由的仙田满。这是因为，他始终将他的儿童游戏环境的研究贯彻于建筑设计中的关系，听到他的名字，就会让人想起那些形态各异的游乐用具，或者是这些设计概念的放大体，即带有儿童游戏和学习环境的复合建筑，从幼儿园到住宅都以同一种概念去理解。以这样的设计概念工作的建筑家，可能不会有第二位，就算有也是儿童方面的定向专门研究。仙田满的研究相对于这种专门化来说，带有异化的性质。作为多元设计活动的一项来说，为儿童设计家具和建筑的人当然是有的，仙田满虽然是以儿童作为构思的起源，但又不局限于此，其最大的差异在于这个思考流程是多样展开的，这与其他的建筑家有很大区别。

据我所知，他是在1960年末，经过菊竹清训建筑事务所的实践锻炼之后独立开业的，创办环境设计研究所的最初，就设计了类似建筑物的游戏用具和带有游戏用具的住宅。从公司的名称上来看，也并不是仅限于建筑单体，同时对建筑环境设计也抱有强烈的兴趣，只有这样才能符合“儿童游戏环境”这一明快的视点，并将设计对象扩展到包含建筑以外的更大的范围。

1970年，当时我编辑的建筑杂志可能是有关仙田满最初的作品集，当时还只是几个游戏用具和一个住宅及店铺设计的实际作品，其余的都尚处于初步设计阶段，但已经体现出了对于环境全体认识的明快轮廓。

“我始终认为有四种游戏空间：自然、密所、废墟、开放空间，这四种空间可看作像布朗运动那样在儿童中以惊人的速度相互碰撞”。

在这样的认识下，他使树状的大型游戏道具带上自然野趣和危险，比如像动物骨骼一样的素材，并且让儿童不去避开，而是预感并体验危险，他认为这对于儿童游戏环境是必要的，这同时也是对于那些所谓的儿童公园，供儿童涂鸦的游戏场所的一种批评。

仙田作为建筑家，不是轻易地将儿童作为研究对象。只是想把握住记忆中最深刻的儿童时代。这并不是怀旧，只是因为身处的时代正是绿林，海岸，以及供儿童游戏的都市环境大量消失的时期。

他在出身地横滨市作了绿化地的现状和未来的详细调查。在杂乱无章的郊外住宅毁掉横滨特性的数年之前就做出了预见，并特别提出一定要保留斜面绿化这一有环境构成的重要因素。他因此而设计了系统化的绿化网络，这和他的长管型游戏道具也是吻合的。

1980年之后，他开始着手设计更大规模的自然博物馆、水族馆和提供现场体验的科学馆等具有展示功能的儿童建筑，这些建筑中当然也会出现些不是为儿童准备的设施，但设计的基本理念和当初独立开业时的想法是一致的，还是以儿童游戏环境作为一贯的设计理念。也就是说他以超越儿童成人的年龄区别的立场去设计建筑，这不是象征权力的古代样式建筑，也不是为提高办公效率而设计的办公楼，是以都市恢复生机为契机的建筑，是对于钢筋混凝土森林的城市战略的藐视。现代建筑是为所有人服务的建筑，这是20世纪所有建筑家所一直提倡的。仙田的儿童游戏环境的思考出发点其实是从身边的场所得到的，他是以个人的执着和感性从这些日常的生活中提炼出这些建筑理念。仙田满所有的作品都是在这种执着和感性中读取的，在乐于其中的同时他自己也发现了可以环境复苏的入口，这样的建筑不仅是横滨或日本，有儿童存在的地域和国家都是需要的。

植田実（うえだまこと）　　　編集長　住まいの図書館出版局

仙田満は東京工業大学工学部建築学科教授をここ１０年あまり勤め、多くの研究と建築設計にもたずさわってきた。また今年春まで日本建築学会会長の地位にあって、建築界全体をとりまとめる労にもあたっていた。だから日本の現代建築の中枢をなす人といっていいのだが、私の見るところ、そうした場所からスタンスをとった自由さが彼の本来の姿のように思えるのである。

それは仙田がこどものあそび環境を、彼の研究あるいは建築設計に一貫したテーマとして持ちつづけているからにちがいない。彼の名をきけば誰しもが、チューブ状の構造体でつくられた遊具を思い出す。あるいはそのコンセプトを拡大した、こどものための遊びや学びの複合建築、あるいは保育園、そして住宅まで、結局は同じ範疇のものとして理解している。

このような仕事をしている建築家は、まずほかにはいない。万一いたとしても、それははじめからこども向け施設という、定められた分野の専門デザイナーとして成り立っている。仙田の仕事は、そうした専門化に対してむしろ唱える性格を内包している。またあるいは、多角的な設計活動の一項として、こどものための家具や建築を手掛ける人もいるのは当然だが、仙田のばあい、発想の源がこどもを起点として、すべてがこの流れにそって多様に展開してきたことが特異であり、ほかの建築家から際だっているのである。

私が知る限り、彼は1960年代の終わり、それまで菊竹清訓の下で建築実務を経験してきた時代から独立して、環境デザイン研究所を設立したときからすでに、建築のような遊具、あるいは玩具箱になぞらえた住宅を設計しはじめていた。またそのファーム名にも端的に窺えるように建築ばかりではなく、環境デザインにも強い関心を示していたが、それこそがこどもの遊び環境という明快な視点によってとらえられていた、建築を含むより大きな対象だった。

1970年、当時私が編集していた建築雑誌で、おそらく初めての仙田満特集を組んだことがある。まだ、いくつかの遊具と住宅１軒と店舗の仕事が実現していたていどで、あとはすべて計画案の段階だったが、それでも彼の環境全体に対する考えは明快な輪郭をすでに提示していたのである。

「４つの基本的なあそびの空間を、ぼくはいつも意識している。自然、アジト、廃墟、オープンスペース。この４つのスペースがこども達のなかでブラウン運動のようにものすごい速さでぶつかり合っているように思える」

こうした認識のなかで、彼は大きな樹木のような巨大遊具、造りものではない、例えば動物の骨そのもののような遊びの素材、危険を避けるのではなく予知し、それとたたかうあそび環境の必要、等々を的確に指摘している。それはいわゆる児童公園といったもの、演出されたガラクタ遊園やラクガキ遊園といったものに対する批判でさえある。

仙田は建築家として、こどもを安易に対象化することをもっとも警戒した。自分自身のうちに身体化しているこども時代の記憶を死守しようとした。だからといってノスタルジーにふける性向があったのではない。日本から丘陵地帯の緑や海岸線、そしてこどもたちが遊びまわる都市環境が、急速に失われつつ時代にあったからである。

彼は自分が生まれ育った横浜市の緑地の現状と将来を、この時期に詳細に調査している。そして開発スプロールが横浜市の特性を跡形もなく消していく数年先を予見し、よい環境のためにとくに効果的な斜面緑地は絶対に保存されなければならないと警告して、そのために系統化した緑地のネットワーク図を描いている。それは彼のデザインした長い長いチューブ遊具のイメージと重なる。

1980年以降、彼はより大規模な自然博物館や水族館や参加体験型の科学館、展示機能やホールを持つ児童施設を手掛けるようになり、必ずしもこどものための施設ではないものも当然そのなかにはあるが、設計の基本理念は独立した当初の、こどものあそび環境の骨格を組み立てるという考えを一貫していることには変わっていない。すなわち彼は年齢によるこどもと成人との区別を超えたところに建築を描こうとしている。それは権威を象徴化する古代的様式でもなければ、企業業務の合理化、効率化を約束するオフィス的建築でもない。現代の都市に生を取り戻す仕掛けとしての建築であり、そこではビルディング・タイプ別の戦略も等閑視されている。すべての人々の日々の生活のための建築こそ、現代の建築であるとは、20世紀全期を通して、あらゆる建築家によって提唱されつづけてきた。仙田の、こどものあそび環境という出発点はじつに小さな身近な場所ではあったが、彼の強靱な持続力と豊かな感性は、その場所を、次々と手掛ける建築のなかに守り抜いてきたのである。

仙田満のすべての建築は、そうした持続と豊かさのなかで読みとれ、そこで楽しむことを通じて自分自身を、また環境を蘇生させる糸口を見出すことができるのである。そうした建築を必要としているのは横浜だけではない。日本だけの事情でもない。こどもがいるすべての地域、国々が必要としているのである。

Makoto Ueda   Chief Editor, SUMAI Library Publishing Company

Mitsuru Senda has been working for Tokyo Institute of Technology, Department of Architecture as a professor for around ten years, and working on many researches and architectural designs. He was also in the position of president of Architectural Institute of Japan till last spring and working for considering about whole architecture society. Therefore, he should be one of persons in the center of Japanese contemporary architecture, but from my point of view, I think freedom he makes distances from those places is true picture of him.

That must be because he keeps consistent theme of children's play environment in his researches and architectural designs. Once hearing his name, everybody who knows him remembers tube-structured play equipment. Or all designs based on his theme such as complex buildings of play and learning for children, day nurseries, even housing…, people understand them in the same category at all. There is no architect who works like him. If there is anybody, they are recognized as specialized designers in a specific filed from the beginning such as facilities for children. Senda's works seem to take the opposite side from those specializations. Also, as a matter of course, there are some people who work on furniture and architecture for children as one of diverse design works, but his uniqueness is that origins of his design ideas starts at children and all of his designs were diversely developed along this flow, and that is the thing which is different from other architects.

As far as I am concerned, when he established Environment Design Institute at the end of 1960's, after working for Kiyonori Kikutake's architectural design office, he already started designing play structures like architecture and housing like a toy box. As the name of the design firm says, not only he got interests in architecture, but also environmental design strongly, and that means to cover larger subjects including architecture - children's play environment, which he had clear vision.

In 1970, an architectural magazine I was editing at that time published a feature article on 'Mitsuru Senda', perhaps his first one. Although most of his works were at the stages of project except some play structures, one private house and a shop building, but he got clear vision and proposed that designs should cover the whole environment.

"I always consider four basic play spaces; nature, anarchy, ruins, and open spaces. I see these four spaces crush each other on a top speed like Brown Movement in children's minds."

In this thought, he indicates appropriately needs of play environment such as giant play structures like big trees, which are not too much designed, and materials of play like bones of animals, which children are away from danger but foresee or fight with it. That could be thought he looks typical children's parks, which are too much directed junk park and graffiti park, with a critical view.

Senda as an architect, alerted most to typicalize children easily. He tried to protect childhood memories of his own. That does not mean he had nostalgic character, but it was the time when we lost urban environment rapidly where children could play around.

He researched greenery in Yokohama of the present time and future in detail at that time, where he grown up. He foresaw situation which Yokohama's characteristics would disappear in several year time because of urbanization and he made a systematic greenery network map while alerting that we must conserve effective slope greenery areas for good environments. That overlaps images of long long tube play structures he designed.

Since 1980, he started working on larger facilities, such as nature museum, aquarium, hands-on type science museum, children's facilities with exhibits and halls. Although some of them are not always facilities for children, there are no changes in basic design concepts since he established the office, which are to construct bones of children's play environments. That is, he tries to think architecture beyond distinction of ages - children and adults. He does not aim ancient style to symbolize authority or office-type architecture to realize rationalization and effectiveness of industry works. It is architecture as contrivance to bring back life in today's cities, therefore it has no meanings of making strategies in different building types. It has been proposed by many architects that architecture for all people's daily life is architecture in the present time. Senda's start point, children's play environments, was truly small near places, but his tough continuance and enriched sensibility has been protecting those places in architecture he worked on. All architectural design of Mitsuru Senda can be read in those continuance and richness, and can be found solutions for revive yourselves and the environment through enjoyment. A place where those architecture is needed is not just Yokohama. Not just in Japan. All the regions and countries where children exist need them.

前言

面向世界之家的环筑（环境建筑）

仙田满　　环境建筑师
东京工业大学教授

环境设计是关系的设计

环境设计是关系的设计。这里所说的关系包括有3个方面，它们是空间的关系、时间的关系和社会的关系。空间的关系是指与用地、道路、邻接用地之间的关系，它必须考虑与视线以及日影、声音、光线等各种有可能发生的一切现象之间的关系。时间的关系就是指历史。它必须尊重过去的各种传统、传说甚至包括“据说”等历史的关系。社会的关系是指所有体现出来的社会活动的关系，它不仅仅只是人间社会的关系，它同时也必须考虑包括人与生物之间的所有生命社会的关系。因此，在我看来设计环境就是“珍惜所有故事的设计行为”。

环境设计的关键在于多样性

环境设计，关键在于多样性的创造。我们知道一个人只有经历了丰富多样的人生体验才能够成熟，所以必须要接触不同的文化，经受其刺激。同时我们也有义务保护各种生物的生活环境。现代的功能主义将环境设计不断地向单一化和分析化的方向引导，这种趋势是完全错误的，在我看来环境设计是在多样性的基础之上进行的统一性的工作。

环境建筑、环建、环境建筑师

1968年我离开菊竹清训建筑事务所之后，成立了环境设计研究所。事务所的名字之所以没冠以建筑设计事务所，是因为我除了想从事建筑设计之外，还想对游戏装置、公园、城市、室内等所有人间的住居环境进行总体的设计。并且，在阅读了宫胁昭的《植物和人间》(NHKBooKS)之后，将人间环境置换成生命环境。这样就形成了我的综合的视点、关系的视点，即经过环境设计了的建筑环境="环境建筑"。并且将我自己的设计作品群称为"环筑"，同时将我自己的职能称呼从建筑师改称为具有更高综合作用的"环境建筑师"。

从自我到他人

关系是有方向性的。当你开始着手设计环境的时候，出发点就是需要被设计的用地本身。设计本身就是要重新审视包括周围环境在内的用地现状的行为。因此从关心自我到关心他人的矢量是最为重要的。现代社会人们往往将自己封闭起来，并切断一切与他人的关系，因而形成了很多自我完结型的空间设计实例。但是这种类型的空间设计绝对不会将城市整体的景观朝向好的方向引导。因此关注他人也是非常重要的。

从内部到外部

对于建筑的评说往往是从周边进行的。的确，外观可以表现出很多解读建筑的元素。但是，我认为体验一座建筑，对于造访者来说从内往外的视线是极其重要的。因为人间只能依靠双脚来回行走，却不可能像小鸟那样地在天空中自由飞翔。从内到外的视线同样适用于地理学家Appleton对于景观的考察。当我们从隐蔽的内部眺望广阔的外部美景时，心情会变得异常平和。

从过去到未来

我们知道未来是不确定的，但是我们可以对于过去的各种已经确定了的东西进行考察和分析，并以此为基础对未来进行预测和展望，所以我从事研究，并认为计划性的研究对于设计活动是非常重要的，通过捕捉人间和空间之间的本质关系，得到它们之间的关系性，可以为下一步的创造工作提供理论根据。同时，不论是研究工作还是设计工作，提出有预见性的假设都是十分重要的，因为没有假设就不会产生创造，可以说从过去到未来的矢量就是创造的假设。

环筑是面向世界的家

展望世界的家这个概念是我在30年前的关于环筑的核心思想。我出生在横滨市的保土之谷这个拥有峡谷状街道的城区。街道特有的地形地势使生活在这里的人们感到非常的舒服和惬意，因此孩子们都从心里害怕走出这里，但是却又对于山对面的景色充满了好奇之心。所以从孩提时代起我一直都患有自闭症。也许是因为出生环境和自身性格的原因，我对于展望世界怀有强烈的欲望。后来大约是1970年左右我在卡密尔的杂记中偶然发现了“展望世界的家”这句话。随后我就决定将这句话作为我的努力方向和环境建筑的理念。展望世界的家，就是具有从自我到他人、从内部到外部、从过去到未来的视点的环筑。

地球环境·建筑宪章

2000年，以建筑学会为中心的建筑系统的五个团体联合制定了《地球环境·建筑宪章》。我担任起草委员长和运用方针作成工作小组的带头人。该《地球环境·建筑宪章》包括以下5条。

1)为了使建筑能够超越世代成为拥有价值的长期被使用的社会资产，它需要被企划、策划、设计、建设、运用、维护(长寿命)。

2)建筑在实现了与自然环境相协调，与多种生物共生存的同时，也成为了构成良好的社会环境的一部分(自然共生)。

3)要将建筑的终生能源消费保持在最低限度，但却要充分地利用自然能源和不使用能源(节能)。

4)建筑要尽可能地减小环境的负荷并且要尽可能地利用能够回收再利用的资源和材料，要将建筑的终生资源消费降低到最低限度(节能和再利用)。

5)建筑要尊重多样的地方文化和历史，要不断地创造出新的文化，并作为良好的成长环境传递给下一个世代(继承)。

我认为该《地球环境·建筑宪章》意味着环境设计正在从以人为中心转化成为以生命为中心。虽然在这里建筑成为核心词语，但是它表示的是建筑环境的意思。所以如果将它置换成土木构筑物也同样适宜。在这个条文中我要特别强调的是第5条“继承”。为了构筑循环型的社会，我们必须将眼光转向下一个世代孩子们的环境问题上。不论是建筑还是城市在这50年里，难道没有破坏孩子们的环境吗？作为负面影响，围绕孩子们产生的各种问题作为当今日本的社会问题暴露了出来。

孩子们能够健康成长的社会就是与自然共生的社会，这是对光明的未来的保障。我们要经常保持面向世界，面向地球的眼光，设计工作对于我们来说难道不是一件很重要的事情吗？作为一名环境建筑师，我希望承担起这个责任。

环游结构

在从1973年到1982年对游戏玩具、儿童公园、游戏园风景以及游戏场所等的调查中，我发现若干能够促使儿童进行游戏的空间结构特征。我称其为“环游结构”。

适宜玩耍的游戏玩具，其游戏动线是循环的。利用游戏玩具的玩耍比赛，大多数情况下都是所谓的“追赶捉迷藏”。因此它的动线当然就要拥有一条封闭的线。另一方面，在对游戏原风景的调查中我发现了这样的空间结构特征，那就是在城市中，能够成为孩子们进行游戏的场所一般都被选择在能够环绕一周的街区，而且有可以抄绕的近道，也称其为短路，同时这些街区都紧邻孩子们经常聚集的小广场。游戏玩具和城市，这两个共通的游戏结构形成了环游结构的基本原理。

在对游戏玩具、游戏场所、建筑、公园等的调查中发现，循环和短路以及紧邻的小广场等结构，在其循环中，“眩晕”是不可缺少的体验元素。同时具有标志性的高大空间或者场所也是必须的，而且需要空间在整体上具有柔软的循环系统，并提供有若干选择，同时开有孔洞可以随时脱离出去。下面将环游结构的特征归纳为以下7点：

(1)拥有循环的功能。

(2)循环(道路)安全并且富于变化。

(3)在其中有标志性空间、场所。

(4)在其循环(道路)上有体验晕眩的部分。

(5)有近道。

(6)循环(道路)上有广场接壤。

(7)整体由多孔质空间构成。

在这里发表的作品几乎都是遵循环游结构而进行设计的。环游结构作为我们设计环境以及建筑的方法，只是普遍和一般的结构体系而并不具有特殊性。

在此希望通过我们的作品能够验证环游结构的有效性。

はじめに

世界を望む家としての環築

仙田　満　　環境建築家
東京工業大学教授

<環境デザインとは関係のデザイン>
環境デザインとは関係のデザインである。関係は空間的関係、時間的関係、社会的関係の３つがある。空間的な関係とは敷地、道路、隣地との関係、具体的な眺望や、日影・音・光等のそこで発する様々な現象との関係において考えねばならない。時間的な関係とは歴史である。過去の様々な伝統、伝承や「言われ」というようなものまでを含んだ歴史的関係を尊重しなければならない。社会的な関係は、そこで営まれる現在の社会的活動としての関係、それは単に人間社会的な関係だけでなく、人間と生物との関係を含めた生命社会的な関係においても考えねばならない。私は環境をデザインするとは「すでにある物語を大切にするデザイン行動」だと定義している。

<環境デザインにおいて重要なのは多様性>
環境をデザインする時において重要なのは多様性だ。人間は多様な体験を通じて成長する。多様な文化に触れ、刺激される。多様な生物の生きられる環境を護る必要がある。現代的機械主義は画一化と分析化の方向の蔓延を導いている。しかしそれは誤りだ。私にとって環境デザインとは多様性を認め、統合化する作業なのである。

<環境建築、環築、環境建築家>
1968年に私は菊竹清訓建築事務所より独立して、環境デザイン研究所を設立した。その名前を建築設計事務所としなかったのは、建築のみならず、遊具も公園も都市もインテリアも人間の住居環境すべてをトータルにデザインしたいと考えたからである。その後、宮脇昭氏の「植物と人間」（ＮＨＫブックス）を読み、人間環境を生命環境に置き換えた。そしてそのように総合的な視点、関係的な視点でつくられた、すなわち環境デザインされたビルトエンバイロンメント＝「環境建築」と呼び、その中で私自身の作品群を「環築」という名前で呼び、又、私自身建築家という職能から、より高度な総合的な立場を目指して「環境建築家」と自称している。

<自己から他者へ>
関係には方向がある。環境をデザインするということは、その出発点は計画すべき敷地という点である。それをその周辺を含めて、もう一度敷地のありようを考えるという行為である。基本的に自己から他者へのベクトルが重要である。往々にして現代都市や地域において、自己閉鎖的に他者との関係を断ち切った形で、自己完結型に空間を形成してしまう例も見られる。しかしその方向は街全体を美しくする方向ではない。他者へのまなざしこそが重要なのである。

<内から外へ>
建築は往々にして外から説明される。外形は確かにその建築の多くをわかりやすく表出している。しかし、建築の内から外への視線がそこを訪れる人にとっては極めて重要である。人は鳥のように空中を自由に飛ぶことはできない。２本の足で歩き回るしかないのである。内から外への視線は、地理学者アップルトンの景観的な考察にも通ずる。私達は広がりのある美しい外を、シェルターの内から眺める時、心やすらかになる。

<過去から未来へ>
未来は不確かである。過去では様々な問題を確かなものに分析し、考察できる。それに基づいた未来への予測、未来への展望が可能である。だから私は研究を行う。デザインをするために計画的な研究が重要である。人間と空間との関係を実体的に捉え、そこに関係性を見いだす。それが次の創造のよりどころになると考える。研究にも設計にも仮説が重要である。仮説が生まれない所に、創造はない。過去から未来へのベクトルはまさに創造的な仮説にかかっている。

<世界を望む家としての環築>
世界を望む家とは、私が３０年前より描いている環築の中心的なテーマである。私は横浜市保土ヶ谷という谷間状の街道の町に生まれた。その谷間はとても居心地よく、その町から出ていくことが子ども心にとても怖かった。しかし山の向こう側が知りたかった。そして子どもの頃から私はずっと閉所恐怖症でもある。そのような生まれ

と性格からか、私は世界を望みたいという欲望を持っていた。そして１９７０年頃にカミユの手帳の中に「世界を望む家」という言葉を発見した。その後、それが私の目標とすべき環境建築のコンセプトとなった。「世界を望む家」とは、自己から他者へ、内から外へ、過去から未来へのまなざしの環築である。

＜地球環境・建築憲章＞

2000年、建築学会を中心とした建築系５団体によって、地球環境・建築憲章が制定された。私はその起草委員長と運用指針作成ワーキンググループ座長を務めた。地球環境・建築憲章は５つの条文より成っている。

１） 建築は世代を超えて使い続けられる価値ある社会資産となるように、企画・計画・設計・建設・運用・維持される。（長寿命）
２） 建築は自然環境と調和し、多様な生物との共存をはかりながら、良好な社会環境の構成要素として形成される。（自然共生）
３）建築の生涯エネルギー消費は最小限に留められ、自然エネルギーや未利用エネルギーは最大限に活用される。（省エネルギー）
４） 建築は可能な限り環境負荷の小さい、また再利用・再生が可能な資源・材料に基づいて構成され、建築の生涯の資源消費は最小限に留められる。（省資源・循環）
５） 建築は多様な地域の風土・歴史を尊重しつつ新しい文化として創造され、良好な成育環境として次世代に継承される。（継承）

この地球環境・建築憲章はまさに人間中心から生命中心の環境デザインへの移行を意味していると考えられる。ここでは「建築」という言葉が中心となっているが、これはビルトエンバイロンメントという言葉を示している。だから土木構築物に置き換えても当てはまる。この条文でも私は第５番目の＜継承＞という所にこだわった。サスティナブルな循環型社会を築くために、特に次世代の子ども達の環境の問題に目を向けなければならない。建築も都市もこの５０年間、子ども達の環境を圧迫してこなかっただろうか。彼らの空間を解放してきただろうか。そのつけとして、子ども達をめぐる様々な問題が、今、日本の社会問題として露呈している。

子ども達が元気に育つ社会、それはとりもなおさず自然と共生する社会であろう。それは明るい未来を保障する。世界への、地球へのまなざしをつねに持ち、デザインすることが、私達にとって重要なことではないかと思う。私は環境建築家としてその責任を果たしたいと考えている。

＜遊環構造＞

1973年から82年までの遊具、児童公園、あそびの原風景、あそび場の調査等から、こどものあそびが成立するためには空間的、構造的ないくつかの特徴があることに気づいた。それを私は「遊環構造」と名付けた。
ゲームが発生しやすい遊具は、あそびの動線が循環している。遊具におけるゲームは、多くの場合、いわゆる「追いかけごっこ」が基本であるから、その動線が一つのクローズした系を持つのは当然である。一方、あそびの原風景の調査で、町の中でのあそびの展開で、こどもたちがあそび場としていたのは、街区が一回りでき、しかも抜け道があったりというようなショートサーキットがあり、またそれにいつもこどもたちが集まる小さな広場がとりついているというような構造であることを発見した。遊具と都市というこの二つの共通したあそびの構造が、遊環構造の基本となった。
循環とショートサーキット、そしてそれに取り付く小さな広場という構造を、遊具、あそび場、建築、公園と、いろいろと調べていくなかで、その循環の中に「めまい」の体験要素が不可欠ではなかろうかということが分かってきた。またシンボル性の高い空間、あるいは場も必要であること、そして全体がやわらかな循環を持ち、いくつかの選択が許され、またさらに外に抜けることのできるような穴だらけの空間であることが要求されていると考えた。そして、遊環構造の特徴を次の七つの条件として整理した。

(1) 循環機能があること
(2) その循環（道）が安全で変化に富んでいること
(3) その中にシンボル性の高い空間、場があること
(4) その循環に「めまい」を体験できる部分があること
(5) 近道（ショートサーキット）ができること
(6) 循環した広場、小さな広場等が取り付いていること
(7) 全体がポーラスな空間で構成されていること

ここに発表する作品はほとんど遊環構造を意識して設計されたものである。遊環構造は私たちの環境や建築をつくりだす方法であると同時に、特殊なものではなく、一般的にも普遍的なものであると考えている。
私たちの作品によって、遊環構造の有効さが確かめられることを望みたい。

Preface

Environmental Architecture as a Home That Overlooks the World

Mitsuru Senda
Environmental Architect
Professor, Tokyo Institute of Technology

<Environmental Design is Design of Relations>
Environmental design is design of relations. The three relations are the relations of space, time, and society. Space relations are that of the property, street, and neighboring property, in which we must consider the relations of specific views, shadows, sounds, light, and other various phenomenon. Time relations are history. The various traditions and legends, "sayings" and such historical relations must also be respected. Social relations are relations existing within current social activities, which not only are relations in human society, but what must be considered is the society of all life that includes humans and other living creatures. I define environmental design as "caring for an already existing tale in designs".

<Diversity is Important in Environmental Designs>
Diversity is important in environmental designs. People grow through various experiences. They see various cultures and are stimulated. It is necessary to protect an environment in which various creatures can live. Modern mechanical principles lead to the spread of uniform and analytic trends. However, this is a mistake. For me, environmental design is to acknowledge diversity and to integrate.

<Environmental Architecture, Kanchiku (Environment Architecture), Environmental Architect>
In 1968, I left Kikutake Kiyonori Architect Office to independently establish the Environmental Design Laboratory. I did not name it an architectural design office because I wished to design not only architecture, but parks and play structures, cities and interior and everything that is a part of mans' living environment. Later, I read Shokubutsu to Ningen (Plants and Humans) (published by NHK Books) written by Akira Miyawaki, and I replaced human environment with life environment. Thus, I called such created in a total and relative viewpoint, in other words, and environmentally designed "built environment", and within that, I call my work "kanchiku". From the function of my job as an architect, I aim for an even higher level and totalistic position in calling myself an "environmental architect".

<From Myself to Others>
Relations have directions. In designing an environment, the starting point if the planned property. It is the act of considering the situation of the property, including that of its surroundings. Basically, the vector from the "self" to "others" is important. Increasingly in modern cities and regions, examples of creation of self-conclusive space that severs relationships from others can be found. However, such direction is not one that makes a beautiful town. The sight of others is precisely what is important.

<From In to Out>
Architecture is often described from the outside. The outer form surely is a clear expression of much of the architecture. However, the view of the inside to the outside of the architecture is very important for those who visit it. People cannot fly the sky like a bird. We must walk around on two feet. The view of the inside to the outside connects to the geographer Appleton's ideas on viewpoints. When we see a beautifully expanding outdoors from inside a shelter, we feel at peace.

<From the Past to the Future>
The future is uncertain. The past we can analyze and study various problems into something certain. Predictions and views of the future based on that become possible. Therefore I experiment. Planned experiments are essential for design. Grasping the actuality of the relation between man and space, and finding the relationship that exists there. That becomes the foundation for the next creation. A hypothesis is important for experimentation and designing. Where there is no hypothesis, there can be no creation. The vector from the past to the future is clearly dependent on a creative hypothesis.

<Environmental Architecture as a Home That Overlooks the World>
A home that overlooks the world is the central theme of Kanchiku that I have been imagining for 30 years. I was born in Yokohama, or more specifically, Hodogaya, which is a town in a valley along a major road. The valley was very comfortable, and as a child, I was afraid of leaving the valley. However, I also wanted to know the other side of the mountain. Moreover, I was claustrophobic from childhood. It may have been from such a birth and personal traits that I had a desire to see the world. Then, around 1970, I found the words "A Home that Overlooks the World" in the ledger of Camus. After that, it became the concept of the environmental architecture that I aim for. "A House that Overlooks the World" is the Kanchiku of the view from the self to others, from inside to out, and from the past to the future.

<Architectural Charter for a Global Environment>
This year, five architectural organizations, with the Architectural Institute of Japan as the core, established the Architectural

Charter for a Global Environment. I took on the role of the head of the drafting committee and chief of the operational guideline production working group. The Architectural Charter for a Global Environment is composed of five articles.
1) Architecture shall be planned, designed, built, operated and maintained as a long-term social property, with a life that will span multiple generations. (Longevity)
2) Architecture shall constitute an element of a sound social environment which is in harmony with the natural environment and which co-exists with the diversity of life on the Earth. (Symbiosis)
3) Architecture shall minimize consumption of energy throughout its lifetime, and maximize the use of natural and unused energy source. (Energy Conservation)
4) Architecture shall incorporate reusable and recyclable resources and materials of minimum environmental loads, and minimize consumption of natural resources throughout its lifetime. (Resource Conservation and Cyclicity)
5) Architecture shall be created as a cultural component, respecting the local history and identity, relating to "genius loci", and succeeding for future generations as a good incubator. (Succession)
The idea is that this Architectural Charter for a Global Environment means the transition from a human-centered environmental design to a life-centered one. Here, the word "architecture" plays a center role, but this indicates the word "built environment". Therefore, it can also be used to mean a constructed object. I was especially particular with the fifth article <Succession>. In order to create a sustainable circulatory society, above all we must look closely at the environmental issues surrounding the children of the next generation. In the past 50 years, both architecture and cities have pressured the environment around children. Have we opened up their space? As a result, the various problems surrounding children is evident in Japanese social issues.
A society in which the children can grow healthily is most probably one in which they can coexist with nature. Such shall guarantee a bright future. I feel that it is important for us to look towards the world and Earth at all times in designing. I strongly wish to fulfill my responsibility as an environmental architect.

<Circular Play System>
Between the years 1973 and 1982, from surveys on play equipment, children's parks, play areas, and memories of childhood play, we came to understand several spatial and structural features that help to establish children's play.
Play equipment that allows game play to develop easily has circulating lines of movement for play. Games using play equipment in most cases take that the basic form of chasing games, so it is natural to have a single running closed system in which movement circulates. On the other hand, according to people's memories of childhood play, in the days when children played in the streets running around the neighborhood, they were able to find shortcuts and a small open space where all children gathered. The common play structure of play equipment and neighborhood forms the basis of my "Circular Play Systems".
In all of my investigations into circulation, shortcuts, and the small open spaces that belong to play equipment structures, play areas, buildings, and parks, I came to realize how essential it is to provide the "thrill experience" in that circulation. Other necessities are a highly symbolic space or area, a gentle overall circulation, several options to choose from, and space with lots of openings to pass through to the outside. I have noted the conditions for play equipment structure features as the following.
1. Has a circulatory function.
2. The circulation path is safe and full of variation.
3. In it there is a highly symbolic space or area.
4. Within the circulation are "thrill experience" sections.
5. Shortcuts can be taken.
6. There is a circular open space, or small plaza, attached.
7. The overall structure is porous.
I have designed children's play environments for nearly 20 years. Around 1982 I extracted these seven conditions from data obtained through surveys, research, and design exercises failures as well as successes. Half my career was complete, therefore, before I ever started thinking about a Circular Play System. Where I had failed, it was largely because there conditions were not fulfilled.
We try to put such verified environmental design theories into practice in our designs and, as mentioned above, refer to the wide range of works created using such theories as "environment and architecture".
Since the office was founded, we have designed many buildings making full use of the two disciplines of research and design. In a way, it is the design which verifies and is actual proof of the interest and reliability of our research.
We shall be delighted, if the examples of design work illustrated in this special issue devoted to Environment and Architecture are proof of the correctness of our way of thinking.

Sports

体育设施

スポーツ施設

SPORTS

从北侧开阔的草坪广场眺望，右侧是中心楼，正面是博物馆楼，并且可以看到设置在其前方的环境探索游戏玩具

北側に広がる芝生グラウンドからみる.右側にセンター棟、正面にドーム棟、その手前に環境発見遊具が見える

View from the expansive lawn on the north side. The Center Block is on the right, the Dome is in the middle, and in the foreground is the Environmental Discovery Play Structures

兵库县立但马球场

所 在 地	兵库县城崎郡
竣工时间	1998 年 10 月
建筑结构	钢结构，部分为钢筋混凝土结构 / 地上 3 层，地下 1 层
用地面积	99134m²
占地面积	19006.58m²(主馆楼) 1380.58m²(中心楼)
建筑面积	21812.77m²(主馆楼) 1139.72m²(中心楼)
获奖经历	1999 年日本建筑学会作品选集入选、2000 年 JIA 环境建筑奖，优良设计奖、日本建筑师联合会奖、HIROBA 奖、2001 年 IOC/IAKS 奖金奖、2003 年 IAA 奖最优秀奖
共同设计	大建设计
结构设备	大建设计、三菱重工业
结构指导	齐藤公男、苫米地司

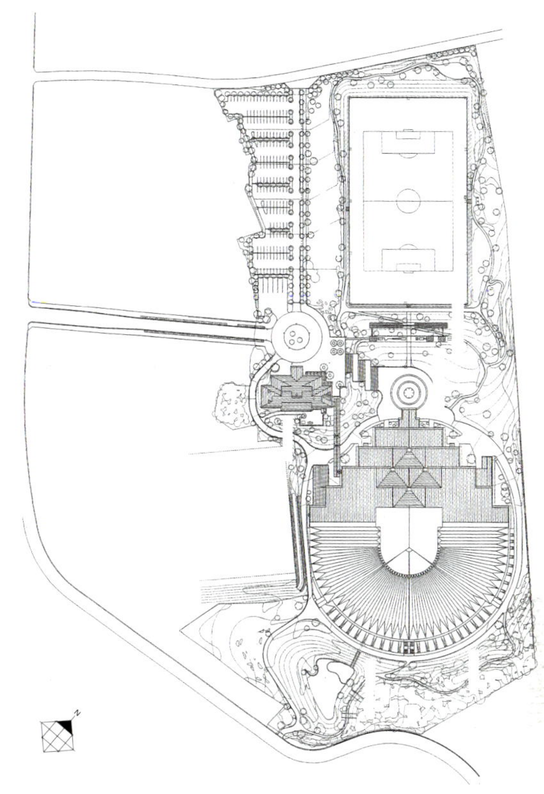

本建筑设计在 1995 年的公开招标中取得了兵库县立但马球场设计竞赛的最优秀奖，建筑形式采用的是可以开闭的建筑形式。在左右 150m，建筑最高高度为 60m 的巨大空间当中，观众席为 3 层构造。从顶层可以欣赏到但马地方的壮丽景观。因为本建筑的设计主题是体育博物馆，所以主馆建筑本身就是集展望、观览、休息、学习等功能为一体的立体环游构造体。建筑外部被开闭部分的聚四氟乙烯纤维膜材料和北侧具有山地建筑风格的多重屋顶的金属材料明显地区分开来，南侧是白色的，北侧是浓茶色的，可以说是一座具有丰富的色彩表情和景观样态的建筑物。

兵庫県立但馬ドーム

所 在 地	兵庫県城崎郡
竣 工 年	1998 年 10 月
規　　模	S 造一部 RC 造／地上 3 階、地下 1 階
敷地面積	99,134m²
建築面積	19,006.58m²（ドーム棟）1380.58m²（センター棟）
延床面積	21,812.77m²（ドーム棟）1139.72m²（センター棟）
受 賞 歴	1999. 日本建築学会作品選集、2000.JIA 環境建築賞、グッドデザイン賞、日本建築士会連合会賞、HIROBA 賞、2001. IOC/IAKS 賞ゴールドメダル、2003. IAA 賞グランプリ
共同設計	大建設計
構造・設備	大建設計、三菱重工業
構造設計指導	斎藤公男、苫米地司

1995年のオープンコンペによって兵庫県但馬ドーム設計コンペ最優秀案となり、採用された開閉ドーム。左右 150m、最高高さ 60m の巨大な空間で、観客席は 3 層構造になっている。上階からは但馬地方の雄大な景観が楽しめる。スポーツミュージアムというコンセプトによってドーム自体が展望、観覧、休息、学習機能を楽しめる立体的な遊環構造を持っている。外観は開閉部分のテフロン膜の部分と北側の山小屋風の多重屋根の金属部分に大きく分けられ、南側は白、北側は渋茶というように色彩にも様々な表情、様々な景観を持つドームである。

Hyogo Prefectural Tajima Dome

Location	Kinosaki-gun, Hyogo Prefecture
Structure	October 1998
Site area	Steel, part RC/1 floor below-grade, 3 above-ground floors
Site area	99,134m²
Building area	19,006.58m² (Dome); 1380.58m² (Center)
Total floor area	21,812.77m² (Dome); 1139.72m² (Center)
Awards	1999 Architectural Institute of Japan Selected Architectural Designs; 2000 JIA Environmental Architecture Prize, Good Design Award, Japan Federation of Architects and Building Engineers Association Prize; HIROBA Award; 2001 IOC/IAKS Award Gold Medal; 2003 IAA Grand Prix
Co-operative Design	Daiken Sekkei
Structural Design	Daiken Sekkei, Mitsubishi Heavy Industries
Structural Design Advisor	Kimio Saito, Tsukasa Tomeji

This retractable dome design was the winner of an open competition held in 1995. It is a huge space, 150m from side to side, and 60m at the highest point, with three tiers of seating for spectators. From the upper level, one has a splendid view of the Tajima area. Created with the concept of a "sports museum", the dome is itself a three-dimensional circular play system where one can enjoy the play, view, rest, or even study. This dome has various faces. The outside appearance has two different scenes, which are Teflon membrane area which can be opened and metal area of multiplex roofs like mountain villa, as well as two colors, white on the south side and dark brown on the north side.

从环境发现游具向中心建筑眺望
環境発見遊具からセンター棟を望む
Viewing the Center Block from the Environmental Discovery Play structures

俯瞰被环绕在但马群山之间的博物馆，用地的约 1/4 是可开闭的博物馆，使广场的土地得到充分地利用

可動屋根を開いたところ.屋根と壁面が一体となって開くためグラウンドレベルまで開放され、遠く但馬の山並みまで望むことができる

The movable roof is opened. People can view mountains of Tajima since the roof and wall open together

场馆2层的连廊，接待处所在的中心栋通向二层的过渡
ドーム2階のブリッジ.受付のある「センター棟」からは2階にアプローチする.
Bridge on the second floor. Visitors reach to this floor in the Dome from the reception in the Center Block through the bridge

场馆3层的体育博物馆
ドーム棟3階のスポーツミュージアム
Sports Museum on the third floor of the Dome

摄影：环境设计研究所
撮影：環境デザイン研究所
Photo: Environment Design Institute

场馆栋和中心栋之间的连接桥
ドーム棟とセンター棟を結ぶ連絡ブリッジ
Connection bridge between the Dome and Center Block

环境探索游戏玩具是 1.8mx1.8m 的组合，全长 57.6rn，分成两列布置
環境発見遊具は 1.8 × 1.8 mのユニットを全長 57.6 m、2 列に配している
The Environment Discovery Play Structures is laid out in 1.8 by 1.8 meter units for a total length of 57.6 meters

北侧夜景

北側夜景

Northern elevation at night

东京辰巳国际游泳馆

所 在 地　东京都江东区
竣工时间　1993 年 3 月
建筑结构　钢筋混凝土结构，部分为钢框架钢筋混凝土结构 / 地上 3 层，地下 2 层
用地面积　22722.35 m^2
占地面积　12319.26m^2
建筑面积　22319.26m^2
获奖经历　1994 年东京建筑奖、照明普及奖、1997 年 IOC/IAKS 银奖
结构设计　结构计画研究所
设备设计　森村设计

本建筑位于东京都临海地区的辰巳临海公园的一角。在设计上要求该游泳馆不仅能够举行重大的国际游泳比赛，同时也可以作为一般市民能够利用的体育娱乐场所。因为在其北侧是运河宽广的水面，因此在空间构成上采取只在南侧看台设置观众席，并将主要游泳池和跳水池的水面与运河水面连接成为一体的设计手法。屋顶通过立体桁架形成以在海面上拍打翅膀的水鸟为意象的屋面形式。在整体设计构思上力求建筑形式与周围环境协调统一，使来到这里的人们，不论是游泳者，还是参观者都能够感受到温馨的环境氛围。

摄影：稻村不二雄
摄影：稻村不二雄
Photo: Fujio Inamura

東京辰巳国際水泳場

所在地	東京都江東区
竣工年	1993年3月
規　模	RC造一部SRC造／地上3階、地下2階
敷地面積	22,722.35m²
建築面積	12,319.26m²
延床面積	22,319.26m²
受賞歴	1994.東京建築賞、照明普及賞、1997.IOC/IAKS賞シルバーメダル
構造設計	構造計画研究所
設備設計	森村設計

東京都の臨海地にある辰巳臨海公園の一角に位置している。国際的な水泳競技大会ができると共に、一般都民のスポーツリクリエーションの場として計画されている。北側に運河の大きな水面があるため観客席を南側のみに設け、メインプールと飛び込みプールの水面と運河の水面が連続するような空間構成とした。海面に羽ばたく水鳥をイメージした5枚のヴォールト状の屋根は立体トラスによって形作られている。全体的に周囲の環境と調和し、泳ぐ人、見る人にとってもやさしい環境となるよう意図した。

Tokyo Tatsumi International Swimming Center

Location	Koutou-ku, Tokyo
Completion	March 1993
Structure	RC, part SRC/2 floors below-grade, 3 above-ground floors
Site area	22,722.35m²
Building area	12,319.26m²
Total floor area	22,319.26m²
Awards	1994 Tokyo Architecture Award, Lighting Diffusion Award, 1997 IOC/IAKS Award Silver Medal
Structural Design	Kozo Keikaku Engineering
Mechanical Design	P.T.Morimura & Associates

Located in Tokyo's Tatsumi Seaside Park. The facility was planned for international swimming events as well as sports recreation of the citizens of Tokyo. The main diving and swimming pools were planned to be a continuation of the open tract of water located on the north side of building and have a consistent water level. Centrally located and at the focus of the five roof-shells designed as an impression of a bird beating its wings on the water. Harmonizing as a whole with the surrounding environment, the intention here was to create an environment which was both swimmer and spectator friendly.

作为二层观众席主入口的甲板
客席のメインアプローチになる２階デッキ
Second floor deck constitutes the main approach to the stands.

内部的泳池和外部的运河水面遥相呼应
内部のプールと外の運河の水面との連続。
The indoor pool is on the same level as the canal outside

强有力的屋面结构在室内展现

ダイナミックな屋根の構成が屋内空間に展開する

The dynamic construction of the roof is echoed in the interior

建筑外观以在海面上拍打翅膀的水鸟为意象
位于中央的红色标志物是胁田爱二郎的作品
海面に羽ばたく水鳥をイメージした外観.中央の動的な赤いシンボルは脇田愛二郎氏作
The exterior, based on the image of a seabird flapping on the water's surface. The dynamic red symbol in the center was designed by Aijiro Wakita

内部的游泳池与室外的运河水面连接形成一体
内部のプールと外の運河の水面との連続
The indoor pool is on the same level as the canal outside

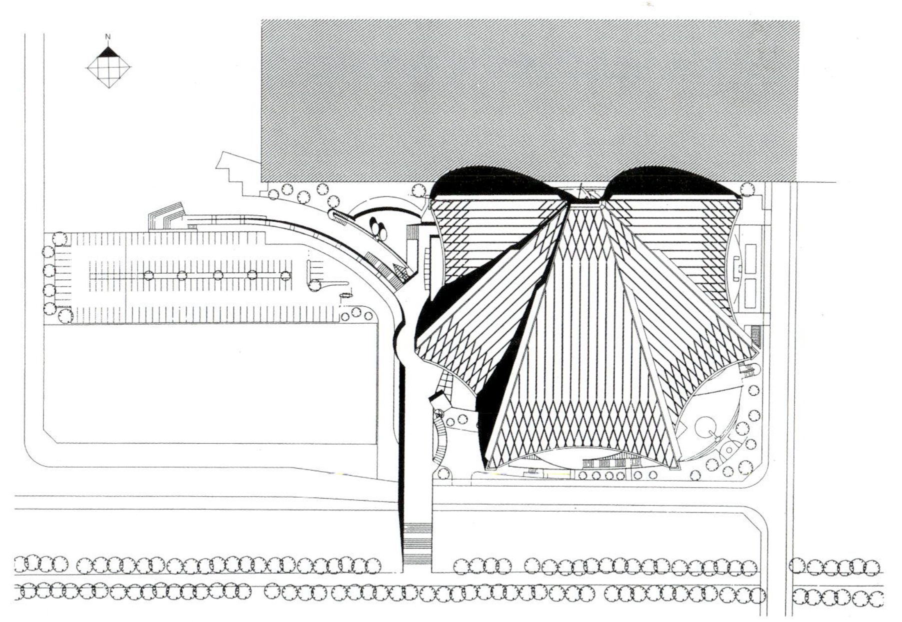

主游泳池和观众席（3500 席）
メインプールと観客席（3500 席）
Main swimming pool and 3500 seats

8 M 100m BTFLY T.R. 4 54.35 NR
SEIKO
1:04.64
1
2
3
4
5
6
7
8
9
10
DS
3 1:07.28
4 1:09.69
2 1:05.21
1 1:04.64

京都亚克体育馆

所 在 地	京都府京都市
竣工时间	2002 年 6 月
建筑结构	钢筋混凝土结构，部分为钢结构 / 地上 3 层，地下 1 层
用地面积	36000m^2
占地面积	7916.97m^2
建筑面积	30586.09m^2
获奖经历	2003 年建筑和社会奖
共同设计	团纪彦建筑设计事务所
结构设计	齐藤公男+结构计画研究所·结构计画普拉斯万
设备设计	叶山成三 + 泰特斯事务所

本建筑是 1998 年在京都市举行的公开招标中被评选为最优秀设计的方案，是国际 B 级公认游泳馆。设施用地位于现有西京极综合运动公园的北侧，跨越阪急电车京都线，用地面积约 3.6ha。除主要游泳场馆和辅助游泳场馆以外，其他建筑物的屋顶部分全部进行绿化处理，一方面可以作为节省能源的对策，另一方面建筑作为公园的一部分，能够使普通市民可以轻松利用，同时试图通过种植栽培植物，实现与自然共生的理念。在设计上注重利用周围环境景观，从主要比赛馆内部可以看到运动公园的绿地以及位于它背后的爱宕山的自然风光，并且在室内、外都配置有水和绿色植物，目的是为了形成能够让使用者以及周围住民都能够享受到自然的舒适环境。

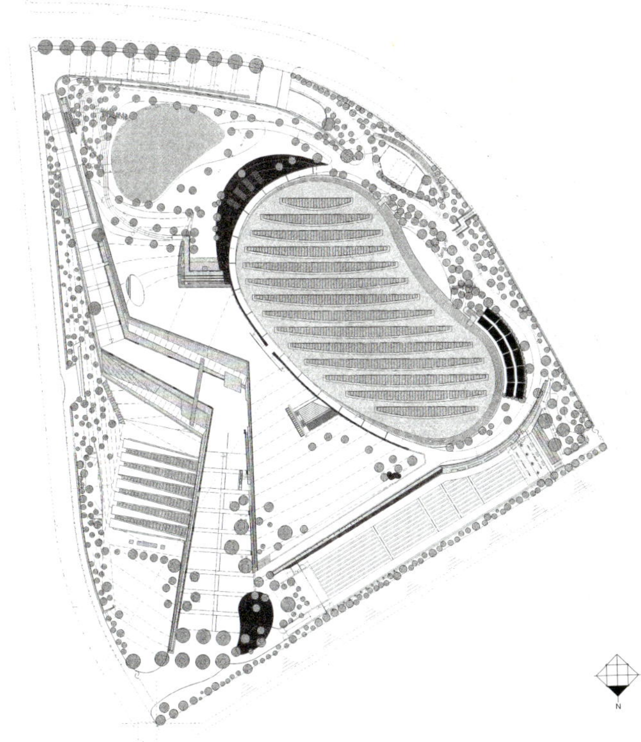

京都アクアリーナ

所 在 地	京都府京都市
竣 工 年	2002 年 6 月
規　　模	RC 造一部 S 造／地上 3 階、地下 1 階
敷地面積	36,000m²
建築面積	7,916.97m²
延床面積	30,586.09m²
受 賞 歴	2003.建築と社会賞
共同設計	團紀彦建築設計事務所
構造設計	斎藤公男＋構造計画研究所・構造計画プラスワン
設備設計	葉山成三＋テーテンス事務所

本施設は、1998年京都市が行った公募型プロポーザルで最優秀案として選定された。敷地は既設西京極総合運動公園の北側、阪急電車京都線を跨いだ3.6haに位置し、国際B級公認プールである。メインプール棟、サブプール棟以外の屋根部分は、緑の丘と称して全面的に緑化し、省エネ対策と同時に、公園部分との一体化による市民利用を図るともに、十分な植栽密度による、建築と自然の共生をめざした。建築計画的にもメインアリーナ内部から運動公園の緑地及びその背後の愛宕山を望む事ができるなど、周辺の環境を考慮しつつ、水、緑を内外に配し、利用者及び周辺住民が自然を享受できる環境形成を意図した。

Kyoto Aquarena

Location	Kyoto City
Completion	June 2002
Structure	RC, part Steel/1 floor below-grade, 3 above-ground floors
Site area	36,000m²
Building area	7,916.97m²
Total floor area	30,586.09m²
Awards	2003 Architecture and Society Award
Co-operative Design	Norihiko Dan and Associates
Structural Design	Kimio Saito + Kozo Keikaku Engineering, Kozo Keikaku Plus One
Mechanical Design	Seizo Hayama + Tetens Engenieering

This facility was chosen as the most excellent proposal in the open competition held by Kyoto City in 1998. With an international B Class approved pool, the 3.6ha site is located on the north side of the existing Nishi Kyogoku General Sports Park, straddling the Hankyu Kyoto train line. In an effort to have encourage usage by the public, through integration with park portions, as well as in striving for coexistence between buildings and nature through ample planting density, roof portions other than main pool wing and sub pool wing, called "green hill," have been entirely covered in greenery for energy saving measures. Consideration had been given to the surrounding environment, and, from the point of view of the construction plan as well, an environment has been created where users and neighborhood residents can enjoy nature through the arrangement of water and greenery indoors and outdoors, as the green sites of the sports park and Mt. Atago behind it can be viewed from inside the main arena.

主游泳馆外观
メインプール棟外観
External view of the main swimming pool building

连接主游泳池和副游泳池的天桥
メインプール棟とサブプール棟を繋ぐブリッジ
Bridge connecting the main and sub swimming pool blocks

主游泳馆入口
メインプール棟入口
Entrance of the main swimming pool building

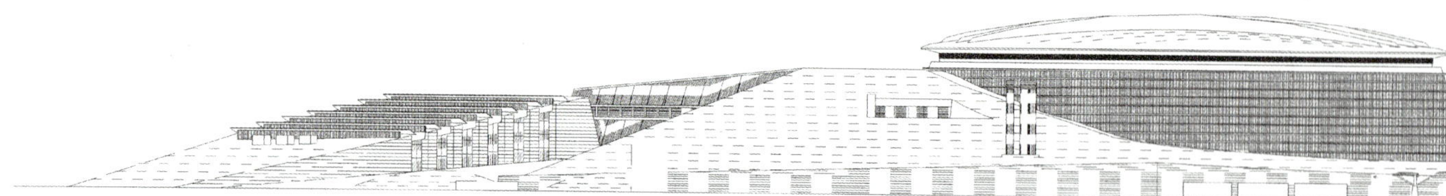

剑术比赛场
アーチェリー練習場
Archery court

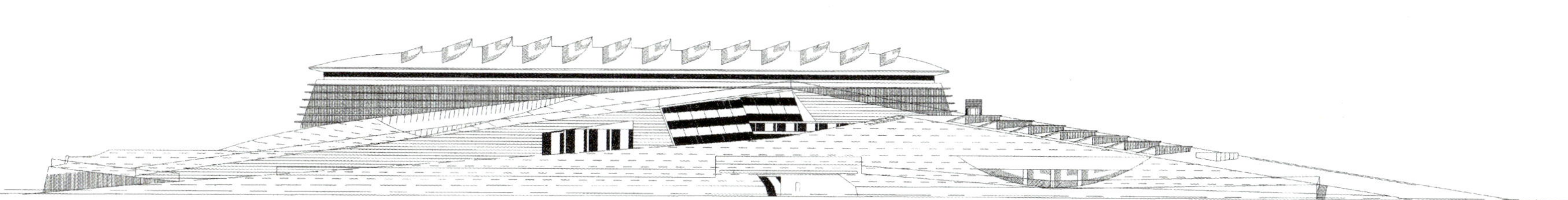

练习泳池
サブプール
Sub swimming pool

观众席
客席
Seating area

从西京极综合运动公园向京都亚克体育馆眺望
西京極総合運動公園から京都アクアリーナを望む
Viewing Kyoto Aquarena from Nishi Kyogoku General Sports Park

2 层走廊

2 階廊下

Corridor on the second floor

主泳池
メインプール
Main swimming pool

夕阳中的正立面
正面夕景
Front elevation at dusk

常滑市体育馆

所 在 地	爱知县常滑市
竣工时间	1993年4月
建筑结构	钢筋混凝土结构，部分为钢框架钢筋混凝土结构/地上4层
用地面积	35158.3m²
占地面积	5099.86m²
建筑面积	9052.72m²
获奖经历	1994年中部建筑奖
结构设计	结构计画研究所
结构设计指导	小野彻郎
设备设计	建筑设备计画研究所

该建筑作为爱知国体的会场将被建在新修建的常滑公园内。主要比赛馆和辅助比赛馆设置在2层，观众席设置在3层，而1层则设置有普通使用者的入口、办公室、更衣室、训练室、会议室等。屋顶的结构形式是立体的钢管桁架，它是由外廊下经过防腐防火处理的钢管组成的斜格子和倒V字形混凝土柱来支撑的，在结构上牢固而简洁。为了体现明快而开放的体育馆风格，建筑整体用半透镜材料所覆盖。

一层大厅

1階ロビー
First floor lobby

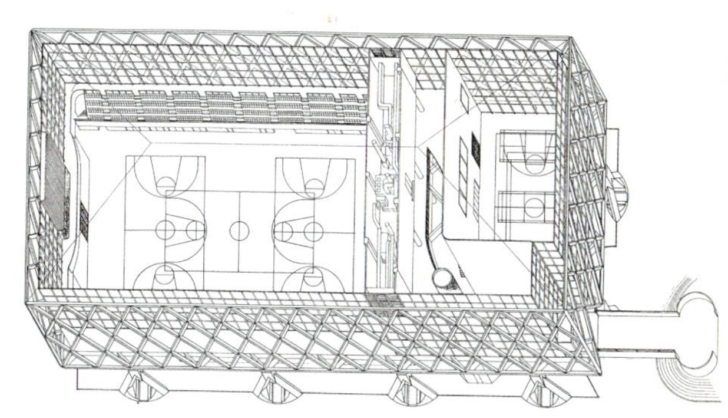

常滑市体育館

所 在 地	愛知県常滑市
竣 工 年	1993 年 4 月
規 模	RC 造一部 SRC 造／地上 4 階
敷地面積	35,158.3m^2
建築面積	5,099.86m^2
延床面積	9,052.72m^2
受 賞 歴	1994.中部建築賞
構造設計	構造計画研究所
構造設計指導	小野徹郎
設備設計	建築設備計画研究所

この施設は、愛知国体の会場として、新設された常滑公園の中に計画された。メインアリーナとサブアリーナを2階に、観客席を3階にし、1階は一般利用者エントランス、事務室、更衣室、トレーニング室、会議室等が入っている。屋根は立体パイプトラスであり、それを外縁で耐候性耐火性鋼管斜め格子と逆V字型コンクリート支柱で支えるというダイナミックでシンプルな構造である。全体をハーフミラーで覆い、明るく開放的な体育館を意図している。

Tokoname Municipal Gymnasium

Location	Tokoname City, Aichi Prefecture
Completion	April 1993
Structure	RC, part SRC/4 above-ground floors
Site area	35,158.3m^2
Building area	5,099.86m^2
Total floor area	9,052.72m^2
Awards	1994 Chubu Architecture Award
Structural Design	Kozo Keikaku Engineering
Mechanical Design Advisor	Tetsuro Ono
Mechanical Design	Kenchiku Setsubi Keikaku Kenkyujo

Planned for the newly constructed Tokoname Park, the venue for the National Athletics Meet in Aichi. There are the main arena and sub-arena on the second floor, seating on the third, and the main entrance, administration room, changing room, training room, meeting room on the first. With a three-dimensional pipe truss roof sustained by a skew lattice of weatherproof and fireproof copper pipes and reverse V shaped concrete at the outer edge, it has a dynamic and simple structure. Completely enveloped in mirrored glass, the intention here was to create a sports facility with a feeling of lightness and openness.

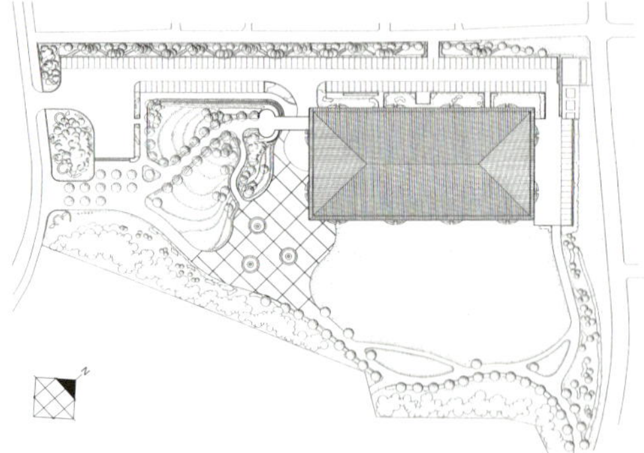

2层大厅
2階ロビー
Second Floor Lobby

主入口大厅
メインエントランスホール
Main entrance hall

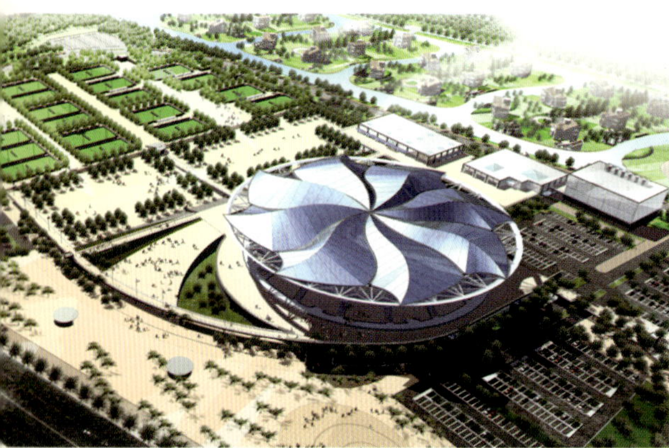

上海旗忠森林体育城国际网球中心项目

所 在 地	中国上海闵行区
竣工时间	2005 年 3 月
建筑结构	钢筋混凝土结构，部分为预制混凝土、钢结构 / 地上 4 层
用地面积	202526m²
占地面积	29567m²
建筑面积	42529m²
共同设计	佐藤尚巳建筑研究所、上海建筑设计研究院
结构设计	结构设计集团
设备设计	综合设备设计、三菱重工

建于中国上海市闵行区的本网球中心，它的占地面积为15ha，建成之后将是与伦敦的温伯尔登，巴黎的劳兰基鲁斯相媲美的世界级网球中心，本设计就是在为此而举行的国际竞赛中被评选为最优秀设计方案的。能容纳 15000 人的本比赛馆，它的屋顶是由 8 枚花瓣形的叶片所组成，8 枚叶片如同照相机的快门一样是可动的。这种可开闭的屋顶结构形式在世界上也属先例。

上海旗忠森林体育城国際テニスセンタープロジェクト

所 在 地	中国上海市閔行区
竣 工 年	2005 年 3 月
規　　模	RC造一部PC造、S造／地上4階
敷地面積	202,526m²
建築面積	29,567m²
延床面積	42,529m²
共同設計	佐藤尚巳建築研究所、上海建築設計研究院
構造設計	構造設計集団
設備設計	総合設備計画、三菱重工業

上海市閔行区に建設される敷地面積15ha、ロンドンのウィンブルドン、パリのローランギャロスに匹敵する世界のテニスセンターが目指され、国際コンペが行われ、本案が最優秀案として選ばれた。スタジアムは15000人収容で8枚の花弁のような屋根がカメラのシャッターのように動く、世界初の新しいタイプの開閉屋根の機構をもつ。

Shanghai Qizhong Forest Sports City Tennis Center Project

Location	Min-hang District, Shanghai, China
Completion	March 2005
Structure	RC, part PC and Steel/4 above-ground floors
Site area	202,526m²
Building area	29,567m²
Total floor area	42,529m²
Co-operative Design	Naomi Sato Architects, Shanghai Institute of Architectural Design and Research
Structural Design	Structural Design Group
Mechanical Design	Sogo Consultants, Mitsubishi Heavy Industries

In an effort to create on a 15ha site in Shanghai's Min-hang District a world-class tennis center competitive with the Wimbledon in London or the Roland Garros in Paris, an international design competition was held and this was the outstanding proposal. With a capacity of 15,000, this stadium has the world's first new-type open-and-close roof structure employing eight flower petal-like roof sections that move like a camera aperture.

浪速综合设施项目

所 在 地　大阪府大阪市浪速区
竣工时间　2005 年 6 月
建筑结构　钢框架钢筋混凝土结构，部分为钢筋混凝土结构、钢结构 / 地上 7 层，地下 1 层，屋顶间 1 层
用地面积　4662.01m²
占地面积　3831.14m²
建筑面积　18263.95m²
结构设计　金箱结构设计事务所
设备设计　综合设备计画

本建筑是建于大阪市中心区难波的大型综合设施，它是由市民体育设施和老人看护养护设施所组成。其中市民体育设施被划分为滑冰比赛馆和大型体育馆区域以及电梯和小型体育馆区域等，并且设置了冰裂缝状的中庭。立面横向的玻璃百叶使得市民设施显得更加开敞明亮和舒适。

浪速複合施設プロジェクト

所 在 地　大阪府大阪市浪速区
竣 工 年　2005 年 6 月
規　　模　SRC造一部RC造、S造／地上7階 地下 1 階、塔屋 1 階
敷地面積　4,662.01m²
建築面積　3,831.14m²
延床面積　18,263.95m²
構造設計　金箱構造設計事務所
設備設計　総合設備計画

大阪市の浪速の都心部に建設される大型市民スポーツ施設と老人デイケア施設の合築である。アイスアリーナと大体育館のブロックとエレベーターと小体育館ブロックに分節化され、その間にクレパス状のアトリウムが設けられている。ガラスの横ルーバーの立面をもち、明るくカジュアルな市民施設とし設計されている。

Naniwa Sports Complex Project

Location　Naniwa-ku, Osaka City
Completion　June 2005
Structure　SRC, part RC and Steel/1 floor be low-grade, 7 above-ground floors
Site area　4,662.01m²
Building area　3,831.14m²
Total floor area　18,263.95m²
Structural Design　Kanebako Kozo Sekkei Jimusho
Mechanical Design　Sogo Consultants

Joint construction of large-scale public sports facilities and elderly day care facilities in urban Osaka's Naniwa area. A crevasse-like atrium was constructed in the division between the section containing an ice arena and a large gym, and the section with a small gym. With vertical plane louvers on the glass side, it is designed as a bright and casual public facility.

Museum

博物馆

博物館

MUSEUM

博物馆公园 茨城县自然博物馆

所 在 地	茨城县岩井市
竣工时间	1994年3月
建筑结构	钢筋混凝土结构，部分为钢结构/地上3层，地下1层
用地面积	163400m²
占地面积	7079.23m²
建筑面积	12771.5m²
获奖经历	1995年展示设计奖、1996年日本造园学会奖、1997年东京建筑奖、1998年建筑百选

该建筑位于南茨城利根川流域的管生池塘边，距离东京有开车一个小时的路程，16.4ha的建筑用地与大约232ha的管生池塘所形成的自然景观非常壮观。在博物馆中设计了形成空间形态结构的"目录空间"，它是由约70m的圆柱柱列所组成，并由被划分为六个主题的展示室形成的柱廊空间所构成。六个主题展示空间通过曲线形的平台连接起来。

ミュージアムパーク 茨城県自然博物館

所 在 地	茨城県岩井市
竣 工 年	1994年3月
規　　模	RC造一部S造／地上3階、地下1階
敷地面積	163,400m²
建築面積	7,079.23m²
延床面積	12,771.5m²
受 賞 歴	1995.ディスプレイデザイン賞、1996.日本造園学会賞、1997.東京建築賞、1998.公共建築百選

東京から車で1時間の南茨城に位置する利根川流域の管生沼のほとりに建設された。敷地は16.4haあるが、約232haの管生沼との一体的な景観が素晴らしい。展示空間は大きく6つに分けられており、それぞれの展示空間は曲線を描くデッキによって結ばれている。これは目次空間と呼ぶ約70mの廊空間として形成された。

Ibaraki Nature Museum

Location	Iwai City, Ibaraki Prefecture
Completion	March 1994
Structure	RC, part Steel/1 floor below-grade, 3 above-ground floors
Site area	163,400m²
Building area	7,079.23m²
Total floor area	12,771.5m²
Awards	1995 Display Design Award; 1996 Japanese Institute of Landscape Architecture Prize; 1997 Tokyo Architecture Award Superior Award; 1998 One Hundred Selected Public Buildings

Located just one hour drive from Tokyo, the museum was built near Sugaonuma in the drainage basin of the Tonegawa. The site is 16.4 hectares, but the scenery as a whole, encompassing the 232-hectares Sugaonuma, is fantastic. There are basically six exhibition spaces, connected by a deck following a curve. This is known as "Contents Space" and forms a corridor of approximately 70m.

主场馆面对充满沼泽植物的湿地，掩映在既存的树林之间

管生沼に面し、既存樹林に身を隠す本館

The main museum building covered by existing woodland, facing Sugao wetland

然

JAN-0940P
然

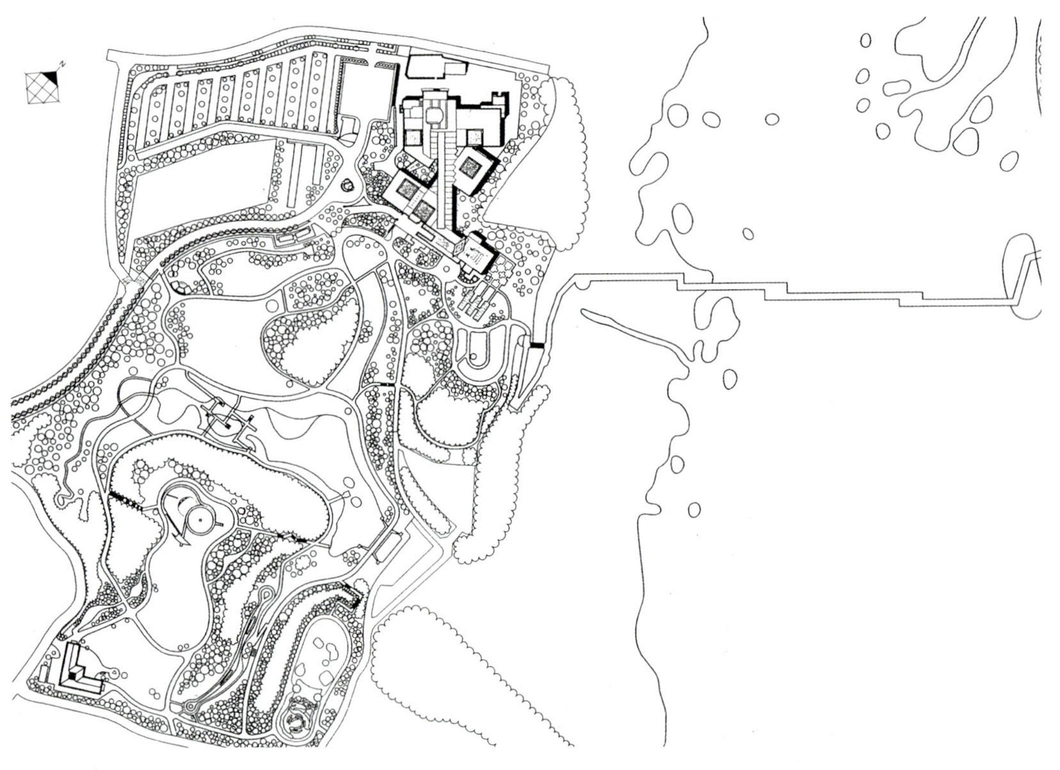

展览空间的二层

目次空間の２階

Exhibition area on the second floor

展览空间的象征性展示——奴奥埃萨鲁斯
目次空間のシンボル展示・ヌオエロサウルス
The symbolic display in the exhibition area, Nuoerozaurus

展厅蜒地球生命史

展示室一地球の生い立ち

Exhibition room – History of earth life

能边观赏充满沼泽植物的湿地，边欣赏乐队演奏的咖啡厅
菅生沼に面し、バードウォッチングが楽しめるカフェ
Cafe in front of Sugao wetland, where people can enjoy bird-watching

1层门厅
目次空間の1階ロビー
Lobby on the first floor

大概 70m 的目录空间

約７０ｍの目次空間
Approximately 70m long exhibition space

由巨大的帐篷和绳网所构成的自然探索器

大きなテントとネットによる自然発見器

This is a Nature Discovery Play Structure made of a large tent and rope netting

大洗水族馆（海底世界）

所 在 地	茨城县东茨城郡
竣工时间	2001 年 10 月
建筑结构	钢框架钢筋混凝土结构，钢筋混凝土结构，部分为钢结构 / 地上 7 层
用地面积	59246.67m^2
占地面积	11620.1m^2
建筑面积	20298.84m^2
获奖经历	2002 年优良设计奖

海底世界大洗水族馆是旧大洗水族馆的改造工程。以往的水族馆大多都是位于海边却设计的非常封闭，与大海相隔绝，而本建筑在设计手法上以“大洗的海约会大海”为主题，将沿着动线景点布置的水槽主线立体地组织到建筑的空间当中，实现了从眼前宽广的大洗的海流经鹿岛海滩、世界的海、水源，最后又回到大洗的海的空间构成，并通过控制和调整空间的明暗变化来营造丰富的空间氛围。

大洗水族館「アクアワールド」

所 在 地	茨城県東茨城郡
竣 工 年	2001 年 10 月
規　 模	SRC 造、RC 造一部 S 造／地上 7 階
敷地面積	59,246.67m^2
建築面積	11,620.1m^2
延床面積	20,298.84m^2
受 賞 歴	2002.グッドデザイン賞

本施設は旧大洗水族館の改築である。従来の水族館の多くが、海辺に立地しながら海に閉ざされているのに対し、アクアワールド大洗水族館は大洗の海＝出会いの海を中心テーマに、動線のストーリーに沿った水槽テーマを建築空間の中に立体的に組み込み、目前に広がる大洗の海から鹿島灘、世界の海、水源をたどり、再び大洗の海へ戻る空間構成を実現させ、それを空間の明暗の変化の演出とも整合させた計画とする。

Oarai Aquarium 'Aqua World'

Location	Higashi Ibaraki-gun, Ibaraki Prefecture
Completion	October 2001
Structure	SRC and RC, part Steel/7 above-ground floors
Site area	59,246.67m^2
Building area	11,620.1m^2
Total floor area	20,298.84m^2
Awards	2002 Good Design Award

This facility is a reconstruction of the old Oarai Aquarium. The theme of 'Aqua World' Oarai Aquarium is centered on an "encounter with the sea" at Oarai, unlike many aquariums up to now which have been closed to the marine environs even though located at the seaside. A composition is implemented in which one proceeds along circular a narrative of tanks, and the theme is inserted dimensionally into the architectural space, and before ones eyes, from the sea at Oarai to Kashimanada, you follow the way along the world's oceans, to the source of water and then again you return to the sea at Oarai--a design that coordinates a performance of changes in the light and shadow of the space.

以渔网和鱼为主题的环游玩具"探索鱼之奥秘教室"

漁網や魚をテーマにした遊具のある「お魚発見教室」

Play structure area with themes of fish and net for fishing, "Fish Discovery Room"

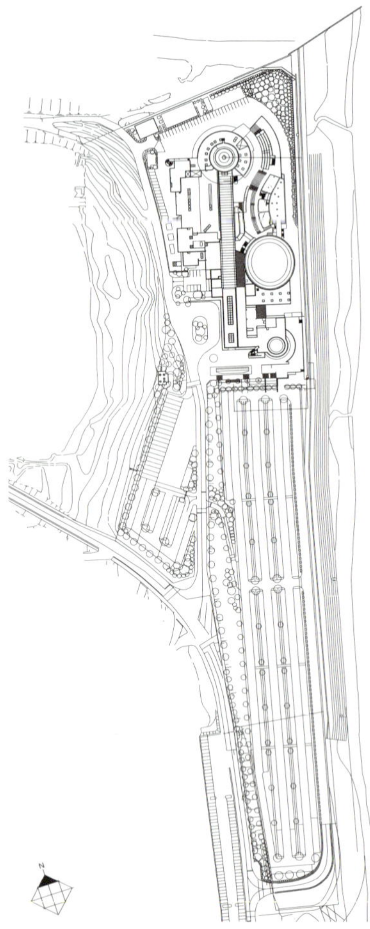

眺望太平洋的中庭
太平洋を望む吹抜
Atrium where visitors can view the Pacific Ocean

大洋之塔的中庭
オーシャンタワーの吹抜
Atrium of Ocean tower

以太平洋为背景观看海豚表演的海洋剧场

太平洋を背景にイルカがジャンプするオーシャンシアター

Dolphins jump up against a background of the Pacific Ocean at "Ocean theater"

从甲板上眺望太平洋

太平洋を望むデッキ
Deck where people can view the Pacific Ocean

户外平台和企鹅游泳池

屋外テラスとペンギンプール

Outside terrace and Penguin pool

关门海峡博物馆 海峡戏剧飞船

所 在 地	福冈县北九州市
竣工时间	2003 年 5 月
建筑结构	钢结构 / 地上 5 层
用地面积	9243.03m²
占地面积	4214.81m²
建筑面积	9898m²
共同设计	福冈县建筑城市部营缮科·建筑设备科、大崎建筑设计事务所、森川哲郎建筑设计事物所

本建筑是作为第二期门司港制动火箭工程的核心设施而建设的，是以历史故事为主题的展示馆，与下关市正在进行的新下关水族馆“海响馆”的建设以及唐户市场整备工程等一同成为关门地区观光开发的重点项目。穿越海峡的船舶以及可以看到对岸景观的玻璃船“戏剧飞船”和符合火箭发射基地特点的“制动火箭”等的综合体形成本建筑的外观形态特征。历史展示部分的内容是通过由国内一流偶人作家的作品来再现历史上的重大事件，因此也希望借此成为偶人博物馆而引起人们的关注。

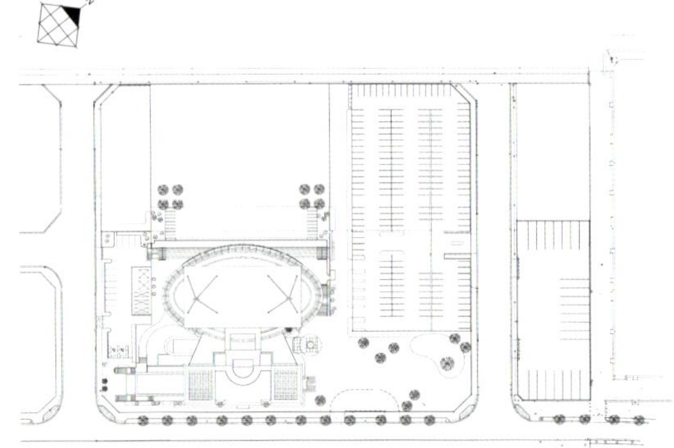

関門海峡ミュージアム 海峡ドラマシップ

所 在 地	福岡県北九州市
竣 工 年	2003 年 5 月
規　　模	S 造／地上 5 階
敷地面積	9,243.03 m²
建築面積	4,214.81 m²
延床面積	9,898 m²
共同設計	福岡県建築都市部営繕課・建築設備課、大崎建築設計事務所、森川哲郎建築設計事務所

本施設は、第 2 期門司港レトロ事業の核施設として計画された歴史ドラマをテーマとする展示館で、新下関水族館「海響館」や唐戸市場の整備を進める下関市と連携する関門観光エリア開発の新拠点と位置付けられている。外観は海峡を通る船舶や対岸からの景観を意識したガラスの船「ドラマシップ」とレトロ地区との連続を意識した「レトロボックス」の複合形態である。歴史展示は国内第一級の人形作家の作品で歴史上のトピックが再現され、人形の博物館としても注目されると期待される。

Kaikyo Dramaship

Location	Kitakyushu City, Fukuoka Prefecture
Completion	May 2003
Structure	Steel/5 above-ground floors
Site area	9,243.03m²
Building area	4,214.81m²
Total floor area	9,898m²
Co-operative Design	Fukuoka Prefecture Depart ment of Architecture and Urban Area, Osaki Kenchiku Sekkei Jimusho, Tetsuro Morikawa Kenchiku Sekkei Jimusho

This facility is an exhibition hall designed on the theme of historical drama as the core facility of the 2nd Moji Port retro revival project, and it's positioned as a new foothold for development of the Kanmon tourist area in co-operation with Shimonoseki City's promotion of the consolidation of the Karato market and the Shinshimonoseki 'Kaikyokan' Aquarium, etc. The exterior is a complex shape that is both a 'Dramaship', a glass boat conscious of things like ships passing through the straits and scenery from the shore and a 'retro box' conscious of a succession of retro districts. As for the historic exhibition, works on historical topics are displayed by one of the country's top doll makers, and it is hoped that it will also receive notice as a doll museum.

夜景，中庭空间、商业街廊将玻璃船、戏剧飞船和象征着门司历史的火箭箱连接成为一体

外観夜景。海側のガラスの船・ドラマシップと門司の歴史を象徴するレトロボックスをアトリウム空間・ガレリアが繋ぐ

Exterior elevation at night. Glass ship by the sea "Dramaship", and "Retro-box" symbolized history of Moji area are connected by atrium "Galleria"

5 层餐厅展望甲板，屋顶的水池如同海峡水面的延续

5階レストラン展望デッキ.屋上の水盤が海峡の水面に連続して見える

Wooden deck outside the fifth floor restaurant. Rooftop pond connects water surface of the sea

从连接门司和下关的联络船上看海峡的戏剧飞船

門司と下関を結ぶ連絡船から見た海峡ドラマシップ

"Kaikyo Dramaship" seen from a ferryboat running between Moji and Shimonoseki

"制动火箭箱"的外观
レトロボックス外観
External view of 'Retro Box'

中庭
アトリウム
Atrium

展示室内部的"海峡火箭大道"
展示室内部「海峡レトロ通り」
Inside the exhibition room, "Kaikyo Retro Street"

海峡中庭

海峡アトリウム
Kaikyo Atrium

由3层结构的螺旋甲板和以大海为意象的网状游戏玩具构成的海峡儿童广场

海峡こども広場.3層構造の螺旋デッキと海中をイメージしたネット遊具で構成している.

Children's plaza. It consists of three-level spiral deck and net play feature with the image of being in the sea

眺望海峡的航拍图

航空写真で海峡を望む

Viewing the Kaikyo by aerial photos

海峡历史回廊内部，通过国内外玩偶作家的作品再现海峡的历史故事

海峡歴史回廊内部.海峡の歴史ドラマを国内外の人形作家の作品で再現している.

Inside "Kaikyo History Corridor". Historical drama at the Kaikyo is reproduced by dolls national and international doll artists created

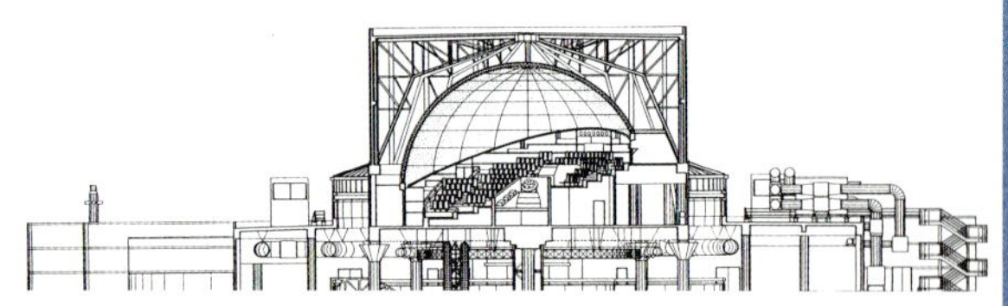

滨松科学馆

所 在 地	静冈县滨松市
竣工时间	1986 年 4 月
建筑结构	钢筋混凝土结构，部分为钢结构 / 地上 4 层
用地面积	9911.47m²
占地面积	3387.32m²
建筑面积	6905.32m²
获奖经历	1988 年滨松城市景观奖，1987 年中部建筑奖，建筑业协会奖

将建筑本身设想为一个大的科学展示物，试图将作为容器的建筑物和作为内容物的科学展示装置融合为一体。在这个建筑中精心设计了玻璃墙的设备室、按颜色划分每个区域、向室内外暴露空调的送风管以及各种配管、电梯及自动门的构造部分、混凝土柱内的配筋、顶棚的龙骨部分等，而且，在外墙壁上设计了根据太阳光的阴影变化可以浮现出科学家面容的装置以及风力发电装置、太阳光追尾装置等，将展示的装置布满建筑全体，进而扩展到包括外部结构在内的整体环境当中。

浜松科学館

所 在 地	静岡県浜松市
竣 工 年	1986 年 4 月
規　　模	RC 造一部 S 造／地上 4 階
敷地面積	9,911.47 m²
建築面積	3,387.32 m²
延床面積	6,905.32 m²
受 賞 歴	1987. 中部建築賞、建築業協会賞、1988. 浜松都市景観賞

建築そのものをひとつの大きな科学展示物と考え、器としての建築と内容物としての科学展示装置の融合を図った。ここではガラス張りの機械室、系統毎に色分けし、内外に露出した空調ダクトや配管類、エレベーターや自動ドアの機構部、コンクリート柱内の配筋、天井下地などを見えるようにするなどの工夫をこらした。また、外壁に太陽光の陰影によって科学者の顔が浮かび出るしかけや、風力発電装置、太陽光追尾装置など、展示的しかけを建築全体に、さらには、外構を含む環境全体にまで及んで設計した。

Hamamatsu Science Museum

Location	Hamamatsu City, Shizuoka Prefecture
Completion	April 1986
Structure	RC, part Steel/4 above-ground floors
Site area	9,911.47m²
Building area	3,387.32m²
Total floor area	6,905.32m²
Awards	1987 Chubu Architecture Award, Building Constructors Society Award; 1988 Hamamatsu Urban Landscape Award

We have advocated the treatment of architecture in these facilities as a large scientific exhibit itself. Here we have tried, through the glassed in machinery room, to make identifiable through the use of different colors air ducts, elevator and automated door machinery, the steel reinforcement of concrete pillars, ceilings and basements, etc. On the exterior walls, the faces of famous scientists are done in a bas-relief visible in the play of sunlight and shadow, and wind-powered electric generators, solar pursuit equipment and other devices are scattered all over the building and beyond, to the outside of the building as well.

科学馆外观
科学館外観
External view of the science museum

浜松科学館

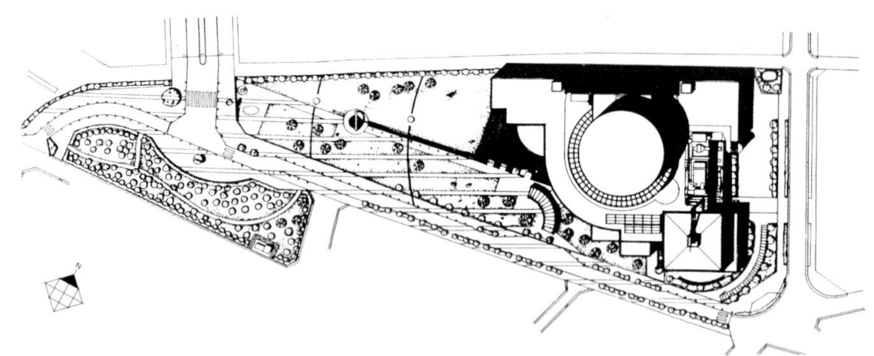

从水的广场眺望科学馆
水の広場から科学館を望む
View the museum from water plaza

馆壁上浮现的科学家肖像
壁面に浮かべるサイエンティストの肖像
The portrait of Einstein appears from the wall

2层展示室"宇宙空间"

2階展示室「宇宙のコーナー」

Exhibition room "The Space" on the second floor

穿过吹拔空间从3层眺望1层

3階から吹き抜けを通して1階を望む

Viewing the first floor from the third floor through hall built in wellhole style

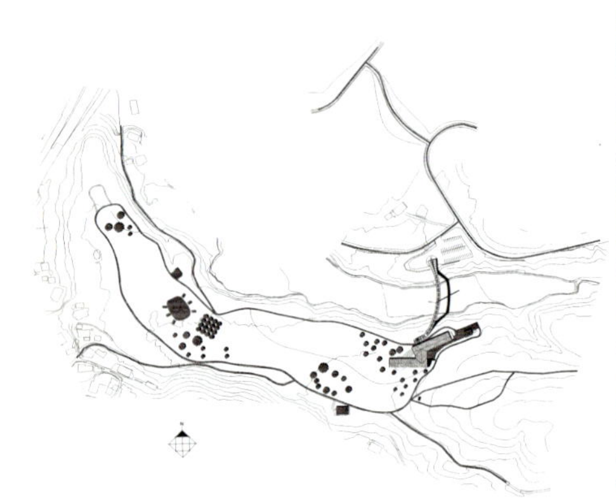

西侧远景，前面是绳文时代住居的复原物
西側からの遠景.手前は縄文屋根住居の復元.
View from the west side. Reproduced Jomon-era house on the front

入口大厅
エントランスホール
Entrance hall.

御所野绳文博物馆

所 在 地	岩手县二户郡
竣工时间	2002 年 3 月
建筑结构	钢筋混凝土结构 / 地上 2 层
用地面积	16104.73m²
占地面积	1719m²
建筑面积	2637.20m²
获奖经历	2000 年岩手景观奖

御所野遗迹被认为是绳文时代中期（大约4500年前）的聚落遗址，是国家级指定历史遗迹。在遗迹公园内摆放有土屋顶的复原住宅，再现了绳文时代的居住环境。其中心设施主要是由介绍该遗迹公园的展示楼和保存收藏迄今为止发掘出的文物的收藏库楼所组成的资料馆。为了使建筑物尽可能地不破坏公园的自然景观，所以在配置上借鉴古代人类的竖坑住居形式，并且在其屋顶上覆盖草坪。

御所野縄文博物館

所 在 地	岩手県二戸郡
竣 工 年	2002 年 3 月
規　　模	RC 造／地上 2 階
敷地面積	16,104.73m²
建築面積	1,719m²
延床面積	2,637.20m²
受 賞 歴	2000.いわて景観賞 （きききのつりはし）

御所野遺跡は縄文時代中期中頃（約4500年前）の集落跡がみられる、国指定史跡である。公園内には土屋根の復元住居が置かれ、縄文時代の環境を再現している。センター施設は主に、この遺跡公園のガイドとなる展示棟と、これまでの発掘資料を保存する収蔵庫棟からなる資料館である。建物は公園からなるべく姿を隠すように配置され、竪穴住居にならい、屋根に芝生を被せた。

Goshono Jomon Museum

Location	Ninohe-gun, Iwate Prefecture
Completion	March 2002
Structure	RC/2 above-ground floors
Site area	16,104.73m²
Building area	1,719m²
Total floor area	2,637.20m²
Awards	2000 Iwate Landscape Award (Bridge 'Kikiki')

The Goshono Ruins are seen as a village ruins from halfway through the middle period Jomon Era (about 4500 years ago), and are a designated national historic ruins. Recreating the environment of the Jomon Era, inside the park are restored dwellings with earthen roofs. The central facility consists mainly of an exhibition that serves as guide to the ruins park, and a reference center composed of storehouse sections for preservation of the materials that have been excavated up to the present. The building is situated so that its shape is concealed from the park, and taking a cue from the pit houses, its roof is covered with grass.

西侧外观，位于建筑物左侧的吊桥的出入口

西側外観.建物左手に吊り橋の出入り口が見える

External view of the building. There is an entrance of "Bridge Kikiki" to the left

吊桥
きききの吊り橋
Bridge Kikiki

吊桥的内部
吊り橋の内観
Inside the bridge

展厅 3
展示室 3
Exhibition room 3

御所野绳文世界
御所野縄文ワールド
Goshono Jomon World

工作间
体験工房
Workshop room

展示室内部
展示室内部
Exhibition room

入口大厅
エントランスホール
Entrance hall

休息角落
休憩コーナー
Rest area

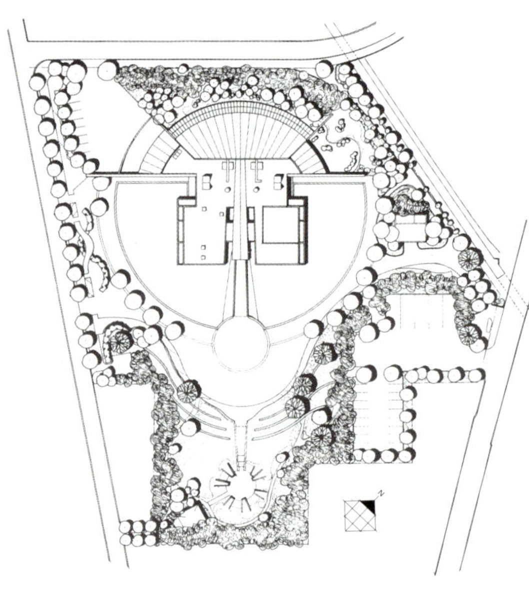

相模川交流科学馆

所 在 地　神奈川县相模原市
竣工时间　1987 年 8 月
建筑结构　钢筋混凝土结构 / 地上 1 层
用地面积　12401m^2
占地面积　1679.07m^2
建筑面积　1662.79m^2
获奖经历　1988年展示设计奖，入选日本建筑学会作品选集，入选神奈川县建筑百选

该建筑是以科学的角度认识相模川为目的而建立起来的淡水水族科学馆。
在设计上将1.2ha的用地作为水系公园融入到设施当中，把弘扬地区的活力和相模川的地域性作为本设计的主要目的。其内部是圆弧形流动着的一体化水槽，通过设置落差，再现了水从相模川的上游流向下游的感觉，并在水中饲养了淡水鱼。

科学馆全貌航拍照片
航空写真.科学館全景
Aerial photo. General view of the River Museum

相模川ふれあい科学館

所 在 地	神奈川県相模原市
竣 工 年	1987 年 8 月
規　　模	RC 造／地上 1 階
敷地面積	12,401m^2
建築面積	1,679.07m^2
延床面積	1,662.79m^2
受 賞 歴	1988.ディスプレイデザイン賞、日本建築学会作品選集、神奈川県建築百選入選

この施設は相模川を科学の目で捕らえた淡水水族科学館である。1.2ha の敷地全体を親水公園として施設に取り込み、地域の活性化と相模川の拠点化を目的に設けられた。内部は円弧状の流れのある一体水槽で、川の上流から川の下流までを落差を設けて再現、淡水魚を飼育している。

Sagamigawa River Museum

Location	Sagamihara City, Kanagawa Prefecture
Completion	August 1987
Structure	RC/1 above-ground floor
Site area	12,401m^2
Building area	1,679.07m^2
Total floor area	1,662.79m^2
Awards	1988 Display Design Award, Architectural Institute of Japan Selected Architectural Designs, Kanagawa One Hundred Selected Buildings

This facility is a museum of fresh-water aquatic life, appropriate to the Sagamiga River. The museum is located on a site of 1.2ha, which includes various facilities in its role as a waterfront park. The museum's main feature is an arched aquarium made to look like a river flowing downstream for the display of freshwater fish.

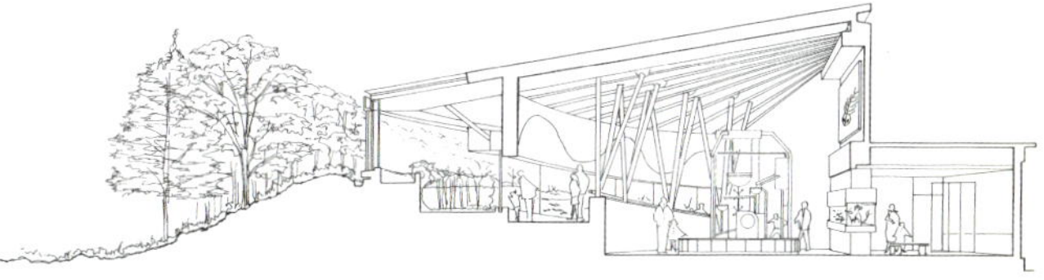

水族馆的流向
流れのアクアリウム
Streaming aquarium

建筑正面的广场和水池
建物正面の広場と池
Plaza and pond in front of the building

航拍照片
航空写真
Aerial view

多摩六都科学馆

所 在 地	东京都田无市
竣工时间	1993 年 12 月
建筑结构	钢筋混凝土结构，部分为钢结构 / 地上 3 层，地下 2 层
用地面积	11083.89m²
占地面积	3269.05m²
建筑面积	6497.86m²
获奖经历	1994 年商业环境设计奖

多摩六都科学馆是由位于东京都北部多摩地区的六个城市共同建设的一座科学馆。科学馆的展示内容设计领域广泛，从宇宙到地区的自然科学无所不及。在空间构成上，为了保留杂木林，将五个展示室控制在三个 10m 的立方体内，并在平面上呈锁链状连接，人们可以透过立方体之间的狭缝观赏到杂木林。为了消除面向杂木林的建筑外墙的存在感，在外墙上使用不锈钢的镜面作为装饰材料。

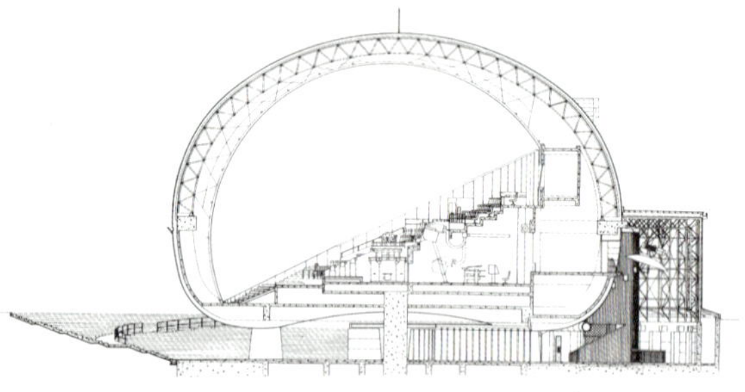

多摩六都科学館

所 在 地	東京都田無市
竣 工 年	1993 年 12 月
規　　模	RC 造一部 S 造／地上 3 階、地下 2 階
敷地面積	11,083.89m^2
建築面積	3,269.05m^2
延床面積	6,497.86m^2
受 賞 歴	1994.商環境デザイン賞

多摩六都科学館は、東京都北部多摩地区の 6 市が共同でつくりあげた科学館である。テーマは、宇宙から地域の自然科学まで多岐にわたっている。雑木林を残すため、5 つの展示室は 3 つの 10m の立方体におさめられ、平面的に鎖状に繋がっている。その立方体の間のスリットから雑木林を楽しむような空間構成としている。雑木林に面した外壁は存在感を消すためにステンレスの鏡面仕上げとしている。

Tama Rokuto Science Museum

Location	Tanashi City, Tokyo
Completion	December 1993
Structure	RC, part Steel/2 floor below-grade, 3 above-ground floors
Site area	11,083.89m^2
Building area	3,269.05m^2
Total floor area	6,497.86m^2
Awards	1994 JCD Award

Tama Rokuto Science Museum was created by six cities of Tama area in northern Tokyo. The display themes broadly cover from the space to natural science in the area. There are five exhibition spaces within the three 10 meter cubes, which are linked on plan, but they have been arranged interestingly so that it is still possible to catch glimpses of the trees outside through slits left between them. The exterior of the building facing the trees is finished in polished stainless steel, in order to reduce its presence.

展厅的休息大厅
展示空間の休憩ロビー
Rest area in the exhibition space

楼梯间，楼梯被称为"儿童电梯"的游戏玩具所围绕
階段室。こどもエレベーターと呼ばれる遊具が階段をとりまく
"Children's Elevator" with the stairs wrapped round it

展厅内部 1
展示室内部 1
Interior of exhibition area1

展厅内部 2
展示室内部 2
Interior of exhibition area2

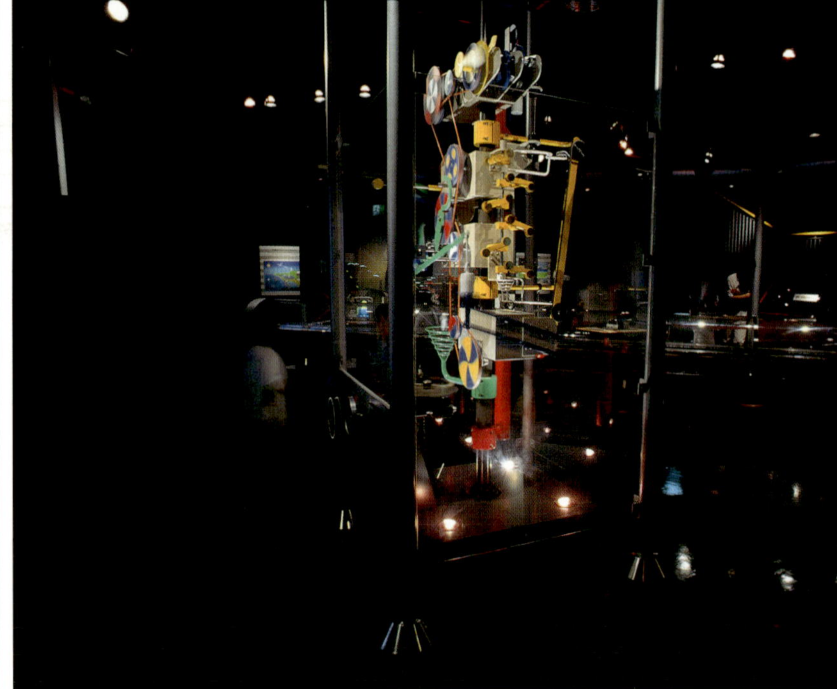

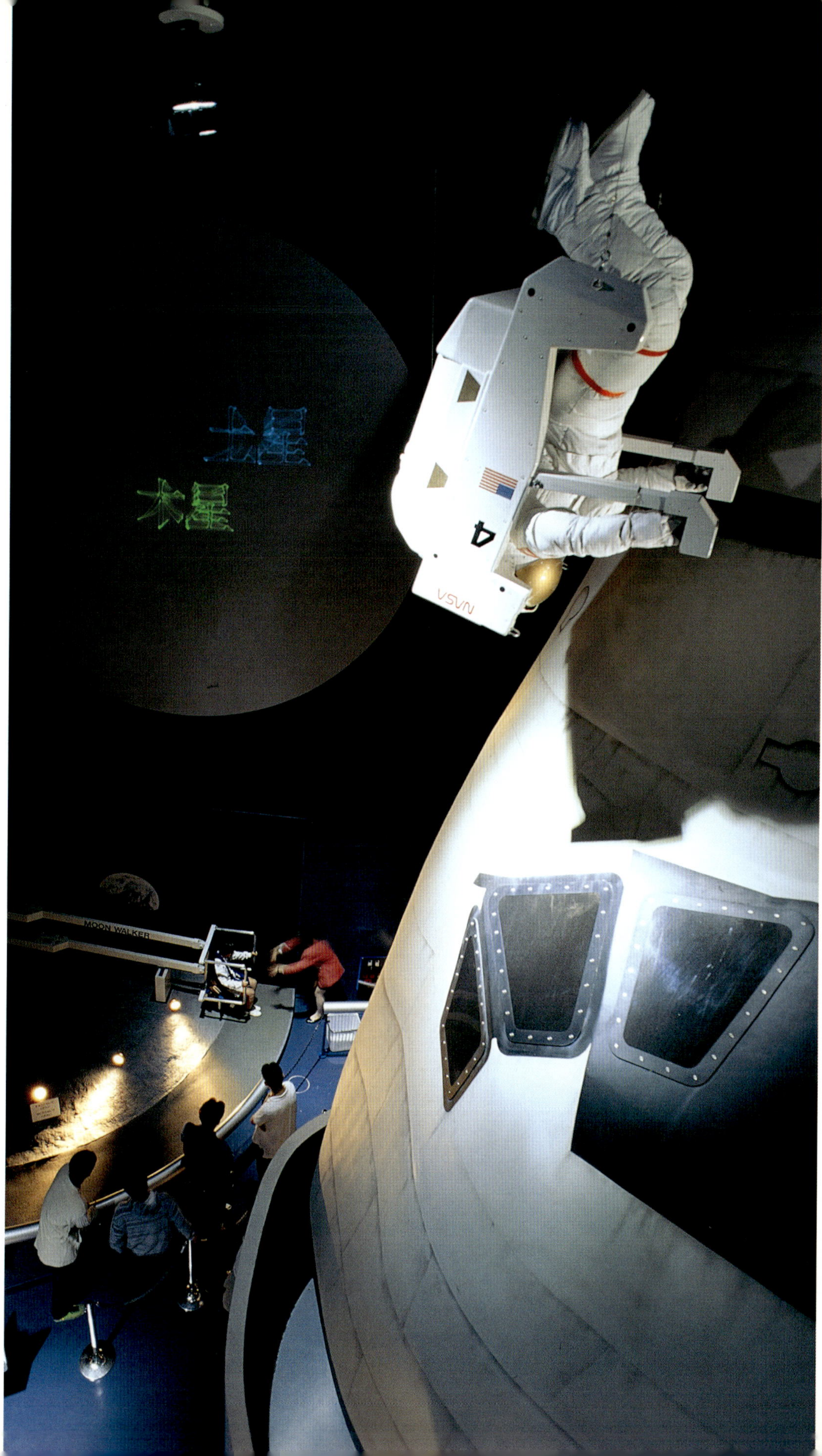

从入口大厅上方向下俯视
エントランスロビーを上から見下ろす
Entrance lobby from above

铃广鱼糕博物馆

所 在 地	神奈川县小田原市
竣工时间	1996年11月
建筑结构	钢筋混凝土结构，钢结构/地上3层，地下1层
用地面积	2624m^2
占地面积	1547m^2
建筑面积	4538m^2

铃广鱼糕博物馆是一座民间的企业博物馆。为了继承传统的工艺技法以及对于手工制作味道的偏爱，将旧的参观工场改造成为可以亲身体验手工制作乐趣的交流场所。在设计手法上通过"鱼·水·板"这三个展示主题，使人们在这里能够得到"制作·观看·娱乐"的生动体验。

鈴廣かまぼこ博物館

所 在 地	神奈川県小田原市
竣工年月	1996年11月
規　　模	RC造・S造／地下1階・地上3階
敷地面積	2,624m^2
建築面積	1,547m^2
延床面積	4,538m^2
受 賞 歴	1997.ディスプレイデザイン賞

鈴廣かまぼこ博物館は民間の企業博物館である。伝統的な技の伝承と手作りの味へのこだわりから、旧見学工場を体験型の手作りコミュニケーションプレイスへと再生した。「魚・水・技」という3つを展示テーマとし、「作る・見る・遊ぶ」体験ができるライブミュージアムを目指した。

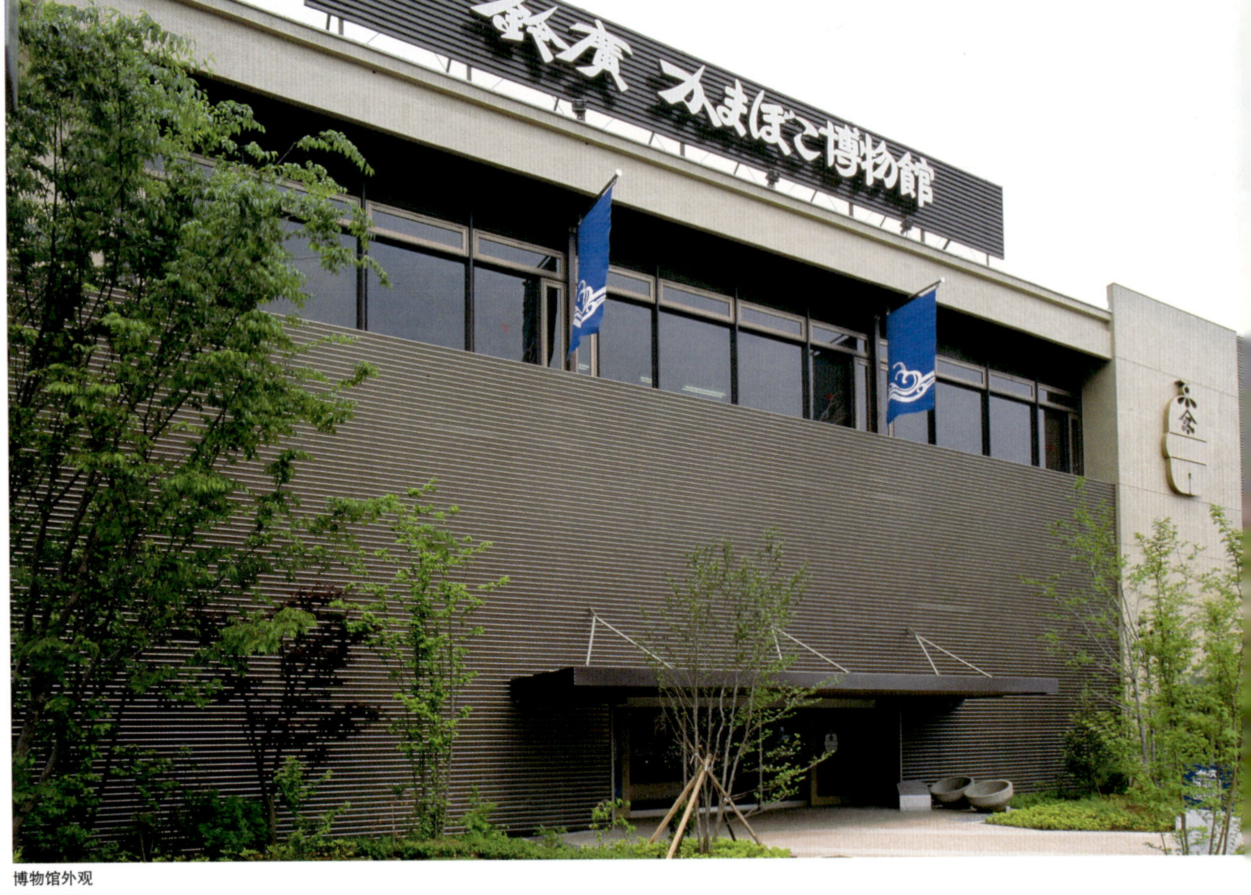

博物馆外观

博物館外観

External view of the museum

Suzuhiro Kamaboko Museum

Location	Odawara City, Kanagawa Prefecture
Completion	November 1996
Structure	RC, Steel/1 floor below-grade, 3 above-ground floors
Site area	2,624m^2
Building area	1,547m^2
Total floor area	4,538m^2
Awards	1997 Display Design Award

The Suzuhiro Kamaboko museum is a private company museum. For transmission of traditional technique and out of the loyalty for handmade flavors, the old factory for visitors has been reborn as a place for communicating experience of the handmade type. There are three exhibition themes, "fish, water and technique," and a lively museum where people can experience "creating, watching and playing".

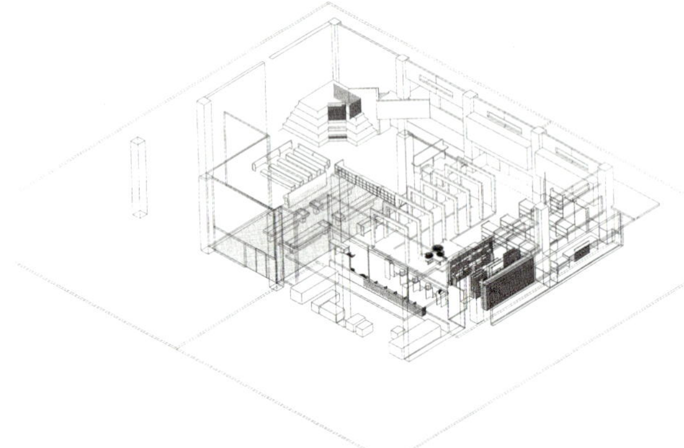

学习操作过程
学習スコープ
Area for learning the process of making Kamaboko

观看“水的回廊”
「水の回廊」を見る
Viewing 'water corridor'

博物馆商店
ミュージアムショップ
Museum shop

展厅内部

展示室内部
Interior of an exhibition room

かまぼこ
手づくり
体験教室

箱根百年水

长崎市科学馆

所 在 地	长崎县长崎市
竣工时间	1996年11月
建筑结构	钢筋混凝土结构,部分为钢结构·木结构/地上4层,地下1层
用地面积	7788m²
占地面积	5053m²
建筑面积	5053m²
获奖经历	1998年度长崎市城市景观奖、入选日本建筑学会作品选集
共同设计	三建设计综合事务所

本科学馆是由外观形态各具特色的展示室、天象仪、天文台等部分所构成，为了使其成为城市的标志物，在设计过程中不断地控制它的体量。展示室的内部以长崎市的街道景观为意象，将空间分为台阶状的六个等级，空间的大小由于远近感的错觉显得更加丰富而有变化，形成独具特色的展示空间。

全貌，建筑的各个部分由不同的形态所构成
全景．建築の各部分が、異なる形態を構成
General view. Each building block consists of different shapes

長崎市科学館

所 在 地	長崎県長崎市
竣工年月	1996年11月
規　　模	RC造・一部S造・木造／地下1階・地上4階
敷地面積	7,788m²
建築面積	5,053m²
延床面積	5,053m²
受 賞 歴	1998.長崎市都市景観賞、日本建築学会作品選集
共同設計	三建設計総合事務所

建築全体のボリュームを抑えつつ、町のシンボルとなり利用者にも分かりやすい施設とするために、展示室・プラネタリウム・天文台の外観をそれぞれ特徴ある形態、仕上げで構成している。展示室内部は長崎市の町並みをイメージし、ひな壇状に6つのレベルに分かれた空間とし、空間の持つ大きさを遠近感の錯覚でより変化させたダイナミックな展示室としている。

Nagasaki Science Museum

Location	Nagasaki City, Nagasaki Prefecture
Completion	November 1996
Structure	RC, part Steel, W/1 floor below-grade, 4 above-ground floors
Site area	7,788m²
Building area	5,053m²
Total floor area	5,053m²
Awards	1998 Nagasaki Urban Landscape Award, Architectural Institute of Japan Selected Architectural Designs
Co-operative Design	Sanken Sekkei Sogo Jimusho

To make this facility into a symbol of the city that is easy for visitors to understand, while holding down the total architectural volume, the exhibition hall, planetarium, and observatory have all been composed with characteristically shaped external appearances and finishes. The exhibition hall has an interior that draws on the cityscape of Nagasaki, and this is divided into the six levels of a hina doll stage, making it a dynamic room where spatial size is altered further by the illusion of perspective.

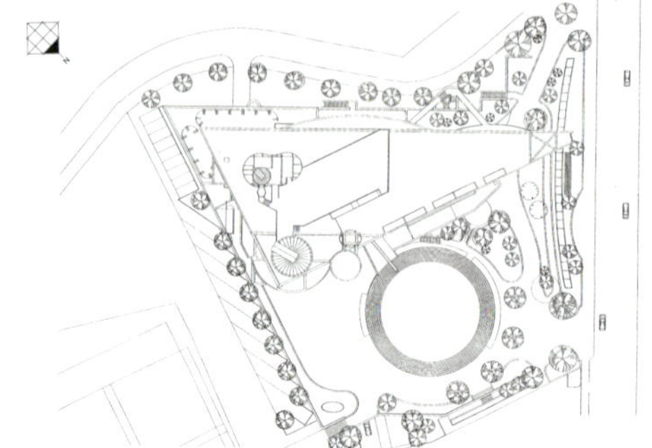

主路夜景
メインアプローチ夜景
Main approach at night

入口
エントランスホール
Entrance

奇异塔内部
ふしぎタワー内部
Kids' elevator (Tower play structure)

山梨县科学馆

所 在 地	山梨县甲府市
竣工时间	1997年12月
建筑结构	钢筋混凝土结构,部分为钢结构·木结构/地上3层,地下1层
用地面积	152386.00m²
占地面积	4725m²
建筑面积	6785.26m²
获奖经历	1999年优良设计奖, 商业环境设计奖

该建筑用地位于自然风景区当中, 曲面形状的建筑沿着用地缓和的等高线进行布置, 并且拥有能够使展示内容依次展开的一体化展示空间。将船的结构引入到建筑当中, 设计了以面向未来实现梦想的"科学之船"为意象的具有象征意义的建筑形态。

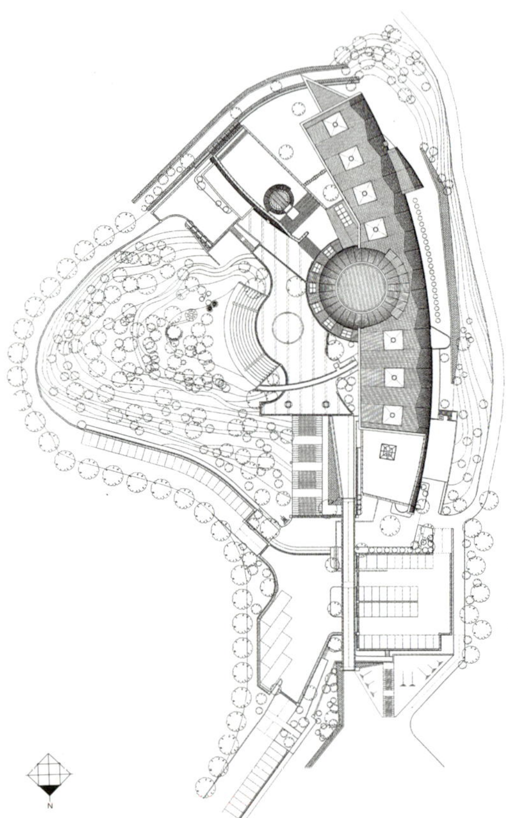

山梨県立科学館

所 在 地	山梨県甲府市
竣工年月	1997年12月
規　模	RC造・一部S造・木造/地下1階・地上3階
敷地面積	152,386.00m²
建築面積	4,725m²
延床面積	6,785.26m²
受 賞 歴	1999.グッドデザイン賞、商環境デザイン賞

風致地区である敷地の等高線に緩やかに沿った曲面形状を持ち、展示ストーリーの効果的な展開ができる一体型の展示空間を持つ。船の構造を建築的に取り入れ、未来へ夢を広げる「科学の船」をイメージさせるシンボリックな形態とした。

Yamanashi Prefectural Science Center

Location	Kofu City, Yamanashi Prefecture
Completion	December 1997
Structure	RC, part Steel, W/1 floor below-grade, 3 above-ground floors
Site area	152,386.00m²
Building area	4,725m²
Total floor area	6,785.26m²
Awards	1999 Good Design Award, JCD Award

With a curved surface that loosely follows the contours of the scenic site, the Center has an integrated exhibition space capable of effectively developing the exhibition narrative. Injecting a ship's structure architecturally, it opens a dream to the future, with a symbolic shape that inspires visualizes a 'Ship of Science'.

森林的万花筒, 映现无尽的树木和天空

森の万華鏡.木や空が無限に映り込む.

"Kaleidoscope of woodland". Trees and the sky reflect on the mirrors infinitely

科学中心全貌

サイエンスシップ全景

General view of the science center

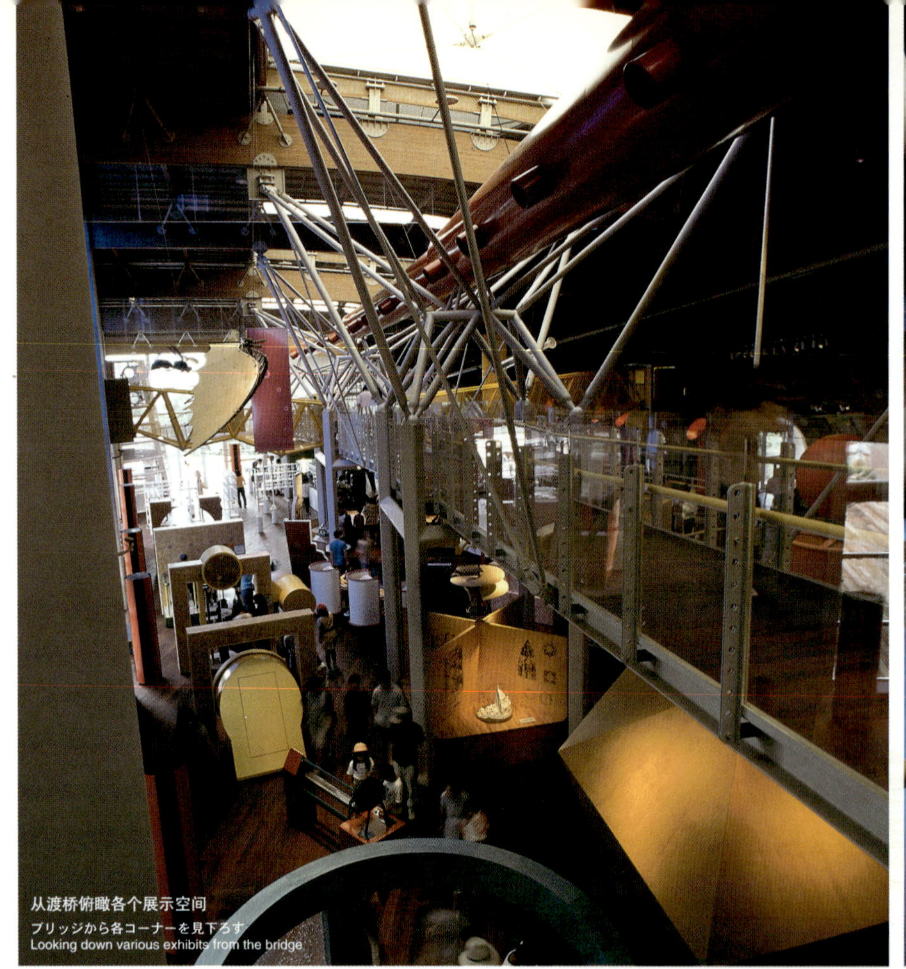
从渡桥俯瞰各个展示空间
ブリッジから各コーナーを見下ろす
Looking down various exhibits from the bridge

桥
ブリッジ
Bridge

室外游戏玩具
屋外遊具
Outdoor play structure

中庭
中庭
Court yard

从展厅仰望
展示空間見上げ
Viewing exhibition space

天桥
空中ブリッジ
Bridge in the air

岐阜县先端科学技术体验中心（科学世界）

所 在 地	岐阜县瑞浪市
竣工时间	1999年3月
建筑结构	钢筋混凝土结构，部分为钢结构 / 地上3层
用地面积	7106.16m²
占地面积	3498.94m²
建筑面积	5970.15m²
获奖经历	2000年中部建筑奖

本设施不同于以往的科学馆，在于它不是以物体展示为目的，而是作为体验型的科学馆来进行策划的，其最大的特点是在馆内不摆放任何展示物品。位于入口门厅上部的讲义实习室的形态参照了离心分离器的形状作为模式。

岐阜県先端科学技術体験センター「サイエンスワールド」

所 在 地	岐阜県瑞浪市
竣工年月	1999年3月
規　模	RC造・一部S造／地上3階
敷地面積	7,106.16m²
建築面積	3,498.94m²
延床面積	5,970.15m²
受 賞 歴	2000.中部建築賞

本施設は、従来の科学館とは異なり展示物を見せるのが目的ではなく体験する科学館として計画され、展示物がないのが最大の特徴である。エントランス上部に位置するレクチャーラボの形態は、遠心分離器をモチーフにしたものである。

Gifu Advanced Science and Technology Experience Center 'Science World'

Location	Mizunami City, Gifu Prefecture
Completion	March 1999
Structure	RC, part Steel/3 above-ground floors
Site area	7,106.16m²
Building area	3,498.94m²
Total floor area	5,970.15m²
Awards	2000 Chubu Architecture Award

Free from any objective to show display items that differ from those in science halls hitherto, this facility was designed as an experiential science hall, and its main feature is the lack of things on display. The shape of the lecture lab, situated in the upper portion of the entrance, has

幕墙夜景
カーテンウォール夜景
View of curtain wall at night

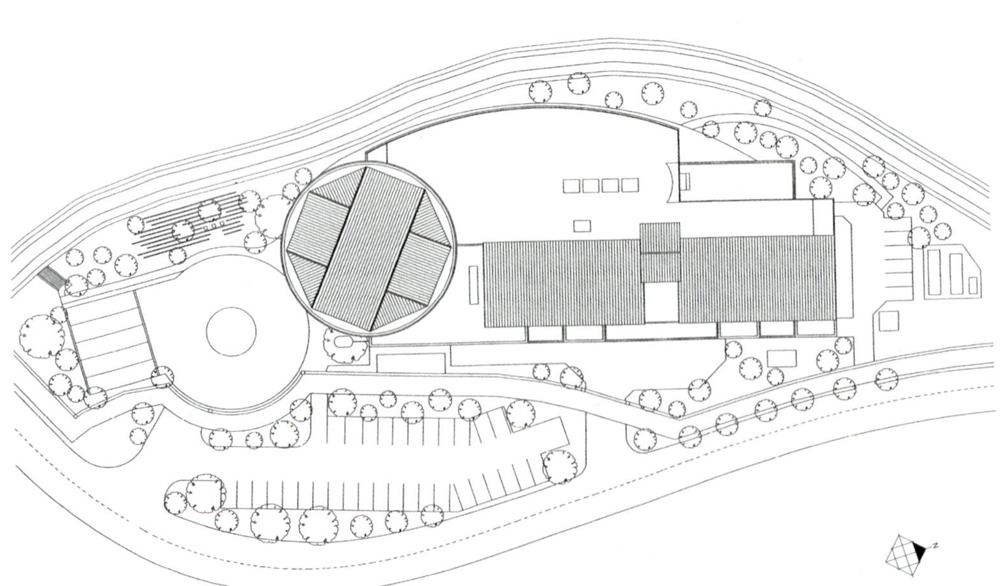

入口上部的倒锥台形状是离心机的形象，里面有一个报告厅

エントランス上部の円錐台の形態は、遠心分離機をイメージ．中には「レクチャーラボ」が入る．

Conic shape above the entrance came from the image of a centrifuge. There is a "Lecture Lab" inside

休息空间
レストコーナー
Resting space

特殊研究室坐席

スペシャルラボ観覧席

Seating area at the special lab

Children

儿童设施

こどもの施設

CHILDREN

富山县儿童未来馆

所 在 地	富山县射水郡
竣工时间	1992年7月
建筑结构	钢筋混凝土结构，部分为钢结构／地上2层，地下1层
用地面积	950000m²
占地面积	2610.57m²
建筑面积	4014.5m²
获奖经历	1993年中部建筑奖、商业环境设计奖、富山县建筑奖、建筑学会北陆建筑文化奖

在位于富山市和高冈市之间的中间位置上有一处95ha的名为太阁山乐园的综合公园，本建筑就建于该公园的中央部，是日本第一座具有美国式儿童博物馆展示功能的儿童设施。建筑平面呈8字形，在功能上共分为3个层面，1层部分为创作活动空间，2层部分为转变了的意见空间，3层部分是作为游戏空间的100m长的回转游戏玩具。在设计上遵循环游结构的原则，即设计出便于孩子们游戏的空间结构，并在屋顶上设置了具有展望机能的屋顶广场。

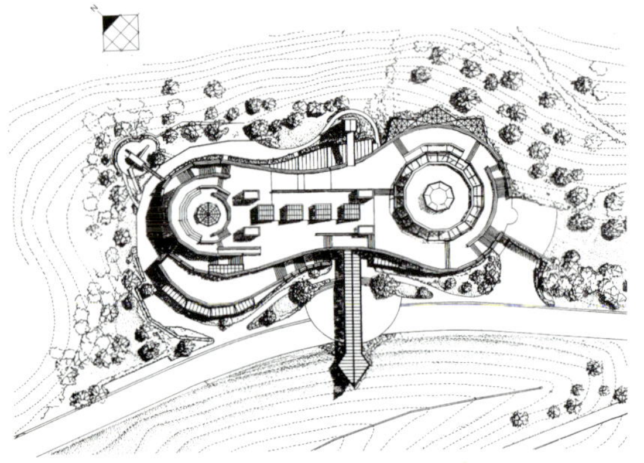

富山県こどもみらい館

所 在 地	富山県射水郡
竣 工 年	1992年7月
規　　模	RC造一部S造／地上2階、地下1階
敷地面積	950,000m²
建築面積	2,610.57m²
延床面積	4,014.5m²
受 賞 歴	1993.中部建築賞、商環境デザイン賞、富山県建築賞、日本建築学会北陸建築文化賞

富山市と高岡市の中間に位置する95haの太閤山ランドという総合公園の中央部に建設された。アメリカのチルドレンズミュージアム型の展示機能を持った日本最初の児童施設である。3層に分かれており、1層部分は創作活動スペース、2階部分は転じた意見スペース、3層部分はあそびのスペースとして100mの遊具が走り回っている。全体として8の字型の平面構成をしており、こども達のあそびやすい空間構成を目指している。屋上も展望機能を持った広場としてつくられている。

Toyama Children's Center

Location	Isui-gun, Toyama Prefecture
Completion	July 1992
Structure	RC, part Steel/1 floor below-grade, 2 above-ground floors
Site area	950,000m²
Building area	2,610.57m²
Total floor area	4,014.5m²
Awards	1993 Chubu Architecture Award, JCD Award, Toyama Architecture Award, AIJ Hokuriku Architectural Culture Award

Located more or less centrally in a comprehensively planned park, this facility combines the kind of "hands-on" exhibition space found in children's museums in the United States with the kind of creative, activity orientated children's facilities common in Japan. Divided into three levels, the first comprises an activity space fostering creativity, the second a "hands-on" exhibition space, and the third a play space where there is a 100 meter long piece of play apparatus. Aiming to create a space in which children would find it easy to play, it describes a figure eight on plan, and then the roof of the building provides the chance to survey the surrounding area.

北侧全貌，钢筋混凝土框架结构的柱间由特殊着色的混凝土砌块双重墙所构成

北側概観.RCラーメン構造の柱間を、特殊着色空間コンクリートブロックの二重壁で仕上げる

View from the north. A double wall of hollow, colored concrete skirts the central framework of the structure

100m长的游戏隧道悬吊在建筑物的顶棚上

建物の天井から100mのプレイトンネルが吊り下げられる

A 100 meter play tunnel is suspended from the ceiling of the building

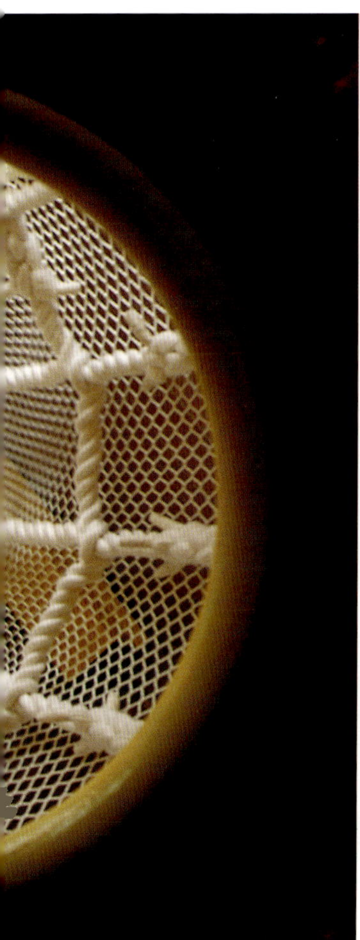

从"结草虫"网向下看，下面乱七八糟
みの虫ネットから見る下界はごちゃごちゃしている
Seen from "Bagworm" net, the world below is a jumping mass

从游戏隧道看外面的世界是五花六色的
プレイトンネルから見る外の景色はさまざまだ
A variety of sights can be seen from the play tunnel

膳食间的聚餐
クッキングルームでの食事会
Kids enjoying what they have cooked together

游玩隧道的入口
プレイトンネルの入り口
The play tunnel entrance

眺望图书角
図書コーナーを望む
Viewing library space

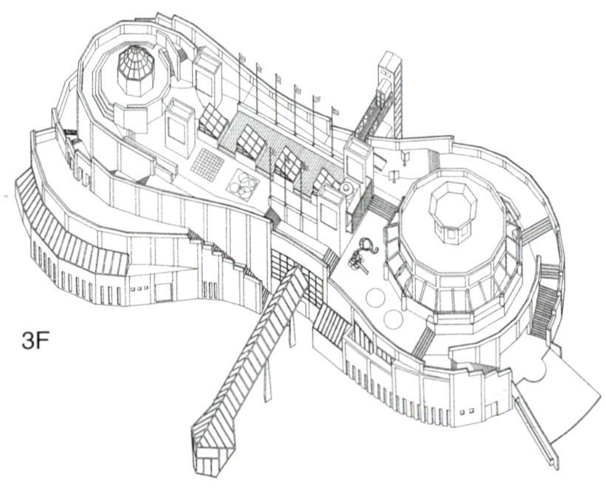
3F

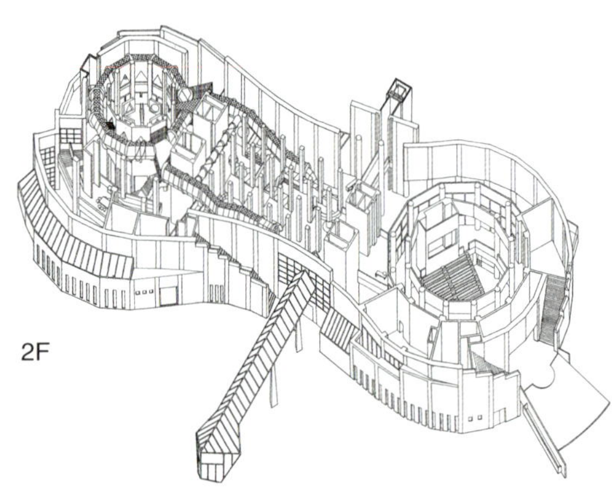
2F

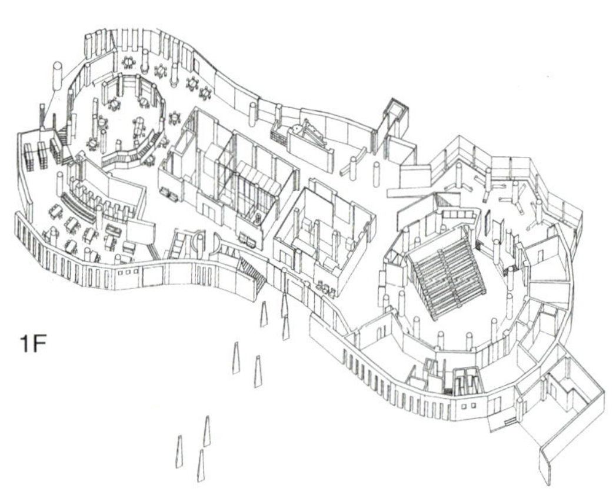
1F

黄色的游玩隧道在空中游走
黄色いプレイトンネルが空中を走っている
The yellow play tunnel running in the air

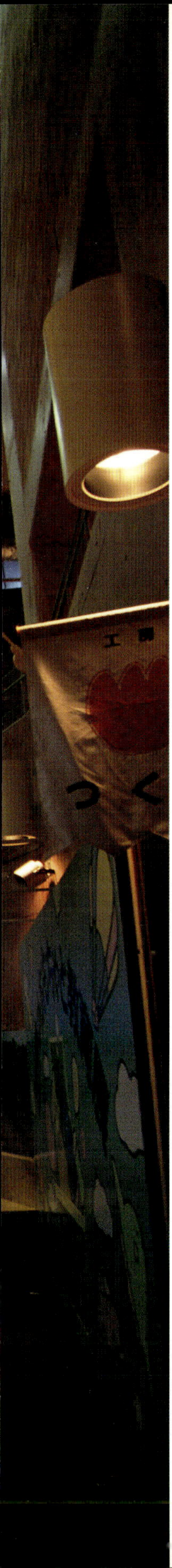

屋顶上的游戏空间
屋上の遊びスペース
Play space on the roof

爱知县儿童综合中心

所 在 地	爱知县爱知郡
竣工时间	1996 年 7 月
建筑结构	钢筋混凝土结构，部分为钢结构 / 地上 3 层，屋顶间 2 层
用地面积	2056.21m^2
占地面积	4673.94m^2
建筑面积	7600.96m^2
获奖经历	1997 年日本建筑学会奖
共同设计	爱知县、藤川原设计

本儿童综合中心建于爱知县长久手青少年公园内，是目前日本最大规模的大型儿童设施。在游戏中庭的上部，除建筑的走廊以外全部都是游戏玩具的回廊。在中央部分是拥有双重螺旋动线的被称为挑战塔的塔状游戏玩具，并支撑着聚四氟乙烯纤维的大屋顶。这本身就是一个具有环游结构的巨大游戏玩具性质的建筑。该建筑不仅是全县的儿童设施，同时也是全县关于儿童游戏环境的研究·学习机构，拥有大的会馆和完善的学习实验设备。

从东侧俯瞰全景
東側上空より見る全景
General view from the east side in the air

愛知県児童総合センター

所 在 地	愛知県愛知郡
竣 工 年	1996年7月
規　　模	RC造一部S造／地上3階、塔屋2階
敷地面積	2,056.21m^2
建築面積	4,673.94m^2
延床面積	7,600.96m^2
受 賞 歴	1997.日本建築学会賞作品賞
共同設計	愛知県、藤川原設計

愛知県長久手青少年公園の中に建設された、日本国内最大規模の大型児童施設である。プレイアトリウム上部には建築的な廊下以外に遊具的廊下が巡らされている。中央部にはﾁｬﾚﾝｼﾞ ﾀﾜｰと呼ぶ二重らせん動線を持つ塔状の遊具建築がテフロンの大屋根を支えている。全体として大きな遊環構造を形成する遊具的建築である。県内の児童施設のセンター機能を持つと同時に、県内のこどものあそび環境の研究・研修機能を持っている。大ホール及び研修実験設備も充実している。

Aichi Children's Center

Location	Aichi-gun, Aichi Prefecture
Completion	July 1996
Structure	RC, part Steel/3 above-ground floors
Site area	2,056.21m^2
Building area	4,673.94m^2
Total floor area	7,600.96m^2
Awards	1997 Architectural Institute of Japan Prize (Architectural Design Division)
Co-operative Design	Aichi Prefecture, Fujikawahara Sekkei

Constructed inside Aichi Prefecture's Nagakute Youth Park, this is Japan's large children's facility. The "play atrium" is encircled at its top part by a corridor for play equipment, in addition to an architectural corridor. At the center part, a tower-like construction of play equipment called the "Challenge Tower" sustains the large Teflon roof with a double helical flow line. Overall, it comprises a play equipment assembly that forms a great play circle structure. As well as possessing the central function of children's facilities inside the prefecture, it functions as a research and training facility for children's play environments. It has also been enhanced with a great hall and training experimental facilities.

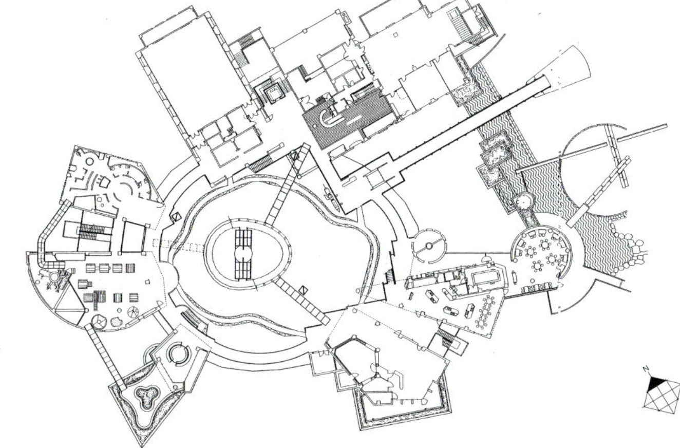

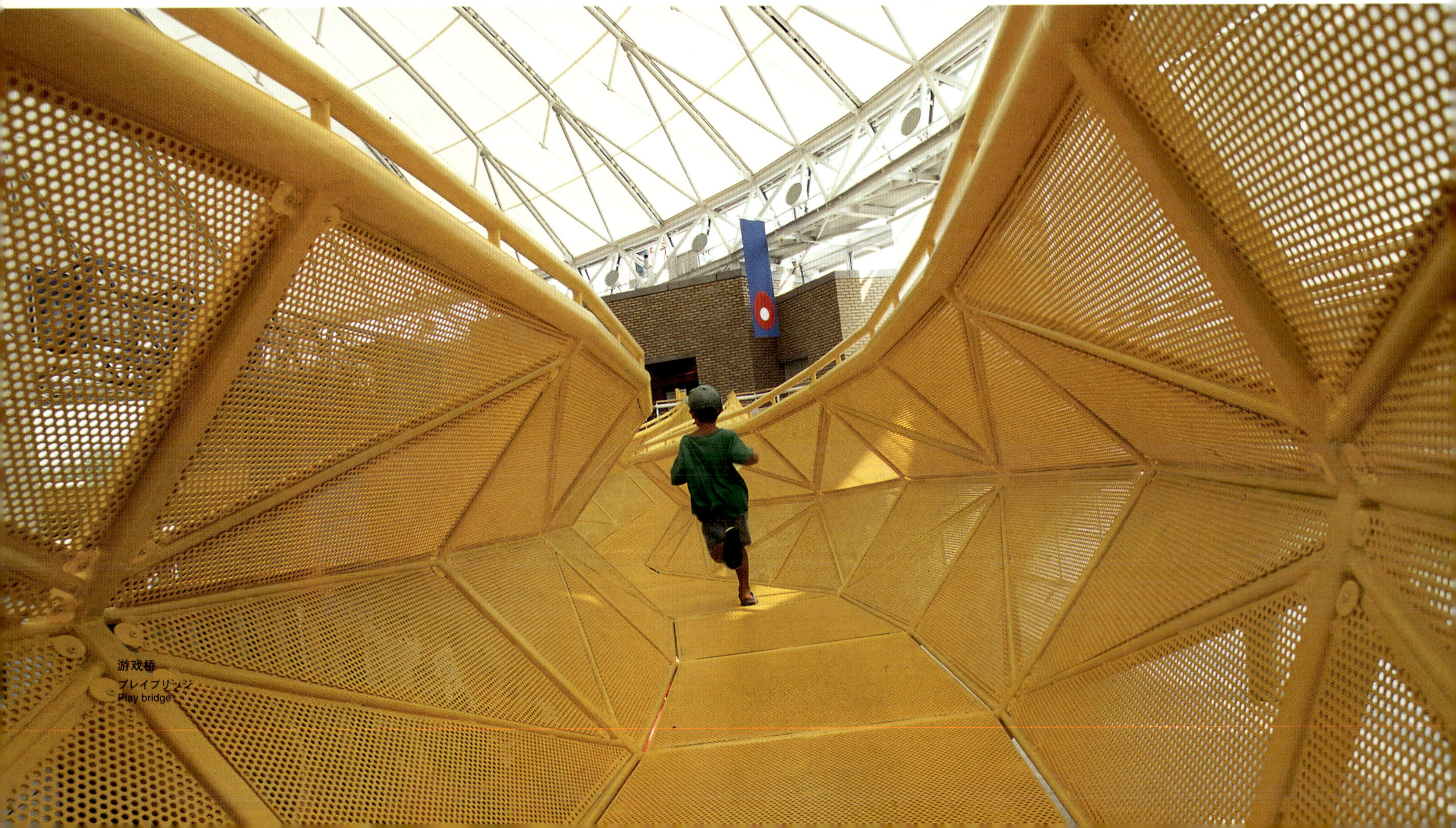

游戏桥

プレイブリッジ

Play bridge

游戏厅内部的“挑战塔”
プレイアトリウム内部の「チャレンジタワー」
Inside the play atrium 'Challenge Tower'

儿童电梯（塔状游戏设施）
こどもエレベーター「塔状遊具」
Kids' elevator of Tower play structure

天桥

空中ブリッジ

Bridge in the air

享受地板材质的乐趣

素材の床を楽しむ

Enjoying the floor material

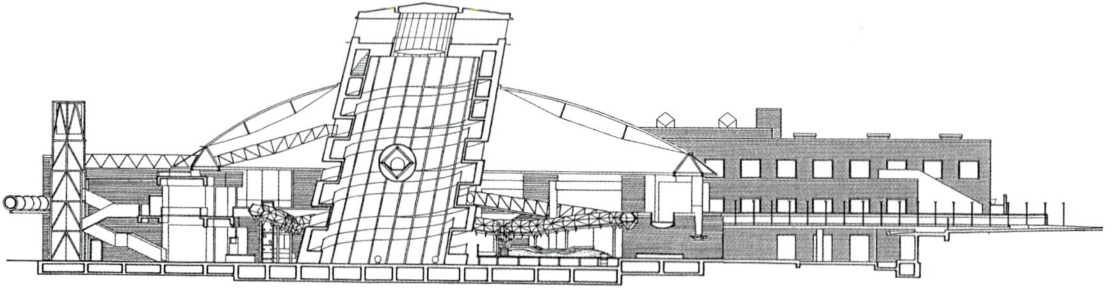

水广场
水の広場
Water plaza.

展望桥

展望ブリッジ
Observatory bridge

"挑战塔"周围的回廊

「チャレンジタワー」の周りの回廊
Corridor around 'The Challenge Tower'

"挑战塔" 内部
「チャレンジタワー」内部
Inside 'The Challenge Tower'

海南市玩吧公园

所 在 地	和歌山县海南市
竣工时间	2000 年 3 月
建筑结构	箱型框架结构，部分为钢结构/地上 2 层，屋顶间 1 层
用地面积	91060.28m^2
占地面积	925.3m^2（风之子馆），414.57m^2（风之回廊·室外卫生间）
建筑面积	1047.63m^2（风之子馆），52.69m^2（风之回廊·室外卫生间）
获奖经历	2000 年故乡建筑景观奖，2001 年景观顾问协会奖优秀奖

风之子馆作为该公园的中心设施，螺旋塔游戏玩具，其主要由两个因素所构成，一个是人们可以一边观看周围的景观，同时还可以顺着坡度上下移动的螺旋状的游戏玩具甲板走动，另外就是遍布其室内的呈螺旋状的网和绳子是各种形式的游戏玩具。

穿过水池遥看风之子馆、风的回廊，回廊全长约 300m
池越しに風の子館、風の回廊を見る.風の回廊の全長は約 300m
Viewing "Kazenoko Tower" and "Outdoor corridor" over the pond
The total length of the corridor is approximately 300m

海南市わんぱく公園

所在地	和歌山県海南市
竣工年	2000 年 3 月
規　模	壁式構造一部 S 造／地上 2 階、塔屋 1 階
敷地面積	91,060.28m²
建築面積	925.3m²（風の子館）414.57m²（風の回廊・屋外トイレ）
延床面積	1,047.63m²（風の子館）52.69m²（風の回廊・屋外トイレ）
受賞歴	2000.ふるさと建築景観賞,2001.ランドスケープコンサルタンツ協会賞優秀賞

この中心施設である風の子館のなかにあるらせん塔遊具は大きく二つの要素から構成されている。一つは風景を見晴らしながらスロープで上がり下りするらせん状の遊具デッキであり、もう一つはその内部に同じくらせん状に張り巡らせたネットやロープをはじめとしたさまざまな遊具アイテムである。

Kainan Wanpaku Park

Location	Kainan City, Wakayama Prefecture
Completion	March 2000
Structure	Steel/2 above-ground floors
Site area	91,060.28m²
Building area	925.3m² (Kazenoko Tower) 414.57m² (Outdoor Corridor, Outdoor Toilet)
Total floor area	1,047.63m²(Kazenoko Tower) 52.69m² (Outdoor Corridor, Outdoor Toilet)
Awards	2000 Wakayama Architecture and Landscape Award; 2001 CLA Award Superior Award

Located on the roof of the children's center, this tower play structure takes on a symbolic aspect while providing a skylight function. The spiral tower is comprised of two major elements. One is the spiral ramp that allows a splendid view while climbing or descending. The other is the interior with its nets, ropes, and other play items in the same spiral.

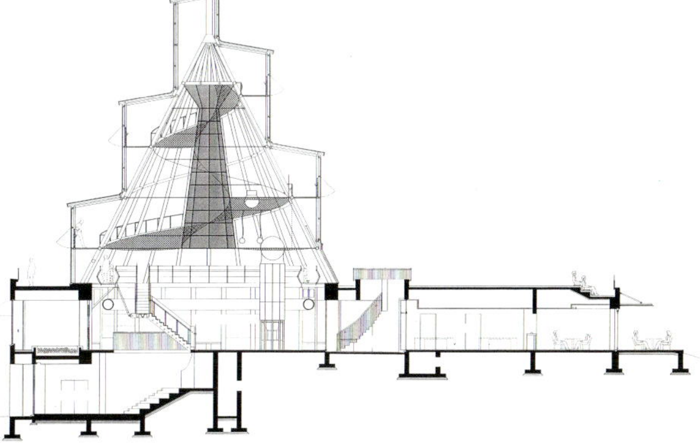

一层的风之子剧场
1階の風の子劇場
Kazenoko theater on the first floor

可在螺旋状网上移动的电梯（塔状游戏设施）

らせん状のネットと上下移動ができるこどもエレベーター（塔状遊具）

Spiral net and Kids' elevator children can go up and down

从草坪广场眺望风之子馆

芝生広場から風の子館を望む

Viewing "Kazenoko Tower" from grass plaza

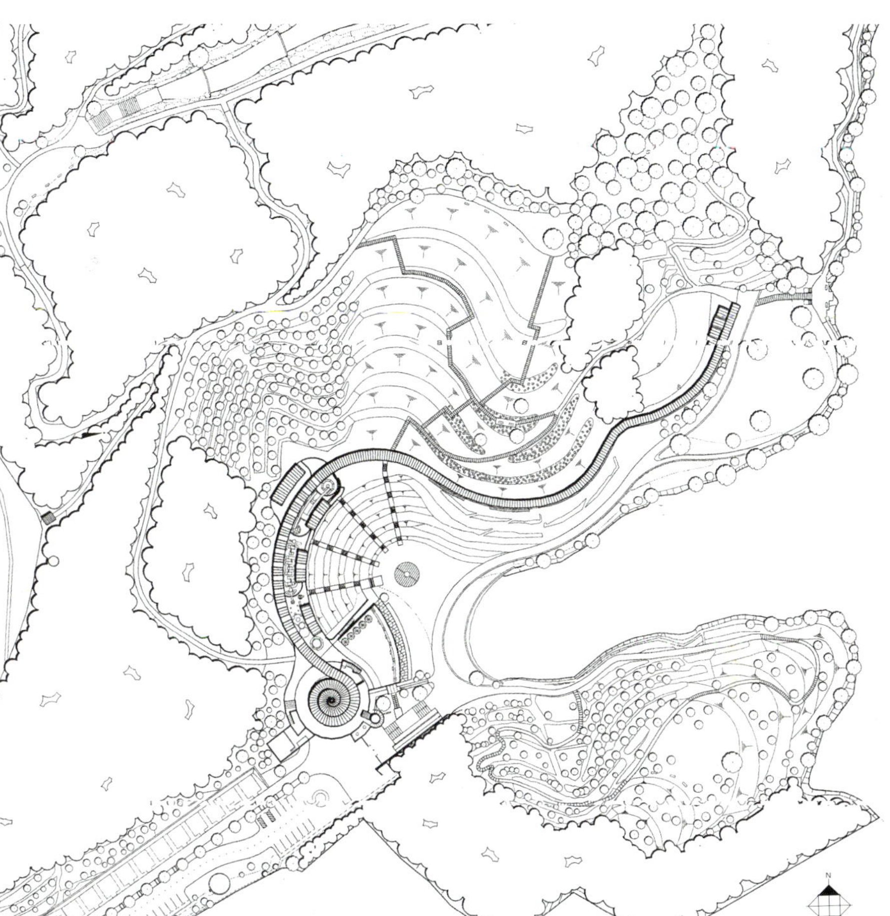
N

風の門

孩子们在借助风之回廊的游戏玩具、台阶，FRP 的管子玩耍的同时，运动技能也可以得到开发

風の回廊の遊具、ステップ . FRP のチューブで、あそびながら運動機能を開発できる.

Play structure by the outdoor corridor, “Step”. They are made of FRP tubes and children can develop their physical skills through play at there

风的回廊

風の回廊

'Outdoor corridor' with roof

南侧外观，中央部分是天文馆楼，上部是天文台，下部是图书馆

南側外観.中央部はプラネタリウム棟、トップは天文台、下部は図書館

The southern exterior. In the center is the planetarium. Above is the observatory and below it the library

伊势原市图书馆・儿童科学馆

所 在 地	神奈川县伊势原市
竣工时间	1988 年 9 月
建筑结构	钢筋混凝土结构，部分为钢框架钢筋混凝土结构/地上4层，地下2层
用地面积	4512.96m^2
占地面积	2921.25m^2
建筑面积	8036.16m^2
获奖经历	1991 年度图书馆建筑奖优秀奖、日本建筑学会作品选集

本建筑是作为与市政府相邻接的图书馆和科学馆的复合设施来进行规划设计的。设计中提出了将图书信息科学馆一体化的概念。具体的方法是，将作为共同部门的信息层布置在2层，图书馆布置在1、2层，而科学馆则设置在2、3层。在现有的市政府、文化会馆之间设计了一个市民广场，将其与图书馆、儿童科学馆之间的通路作为交流道路而与广场融合形成一个整体。

伊勢原市立図書館・子ども科学館

所 在 地	神奈川県伊勢原市
竣 工 年	1988 年 9 月
規　　模	RC 造一部 SRC 造／地上 4 階、地下 2 階
敷地面積	4,512.96m^2
建築面積	2,921.25m^2
延床面積	8,036.16m^2
受 賞 歴	1991. 図書館建築賞優秀賞、日本建築学会作品選集

本施設は市役所に隣接した図書館と科学館の複合施設として計画された。私たちは図書情報科学館という一体的なコンセプトを提案し、具体的には共通項である情報フロアーを2階に配し、図書館を1、2階に、科学館を2、3階とした。既存の市役所・文化会館との間に市民広場を構成するような配置計画とし、図書館・こども科学館との間を抜ける通路をコミュニティ道路とし、広場と一体化した。

Isehara Library and Science Museum

Location	Isehara City, Kanagawa Prefecture
Completion	September 1988
Structure	RC, part SRC/2 floor below-grade, 4 above-ground floors
Site area	4,512.96m^2
Building area	2,921.25m^2
Total floor area	8,036.16m^2
Awards	1991 Library Architecture Award Superior Award, Architectural Institute of Japan Selected Architectural Designs

Although it was planned as a combined Isehara library and science museum on an adjoining site to the city hall, we proposed a comprehensive facility comprising library and information science functions. In concrete, the information area, which is common in both, was provided on the second floor, and the library on the first and second floors, and science museum on the second and the third floors. The site was planned so as to have some open spaces between the existing city hall and cultural hall complex on the adjoining site and the remaining passageway between the two sites was arranged as an integral part of these public spaces.

航空照片
航空写真
Aerial view

图书馆的阅览室

図書館.ブラウジング

Right: Browsing area

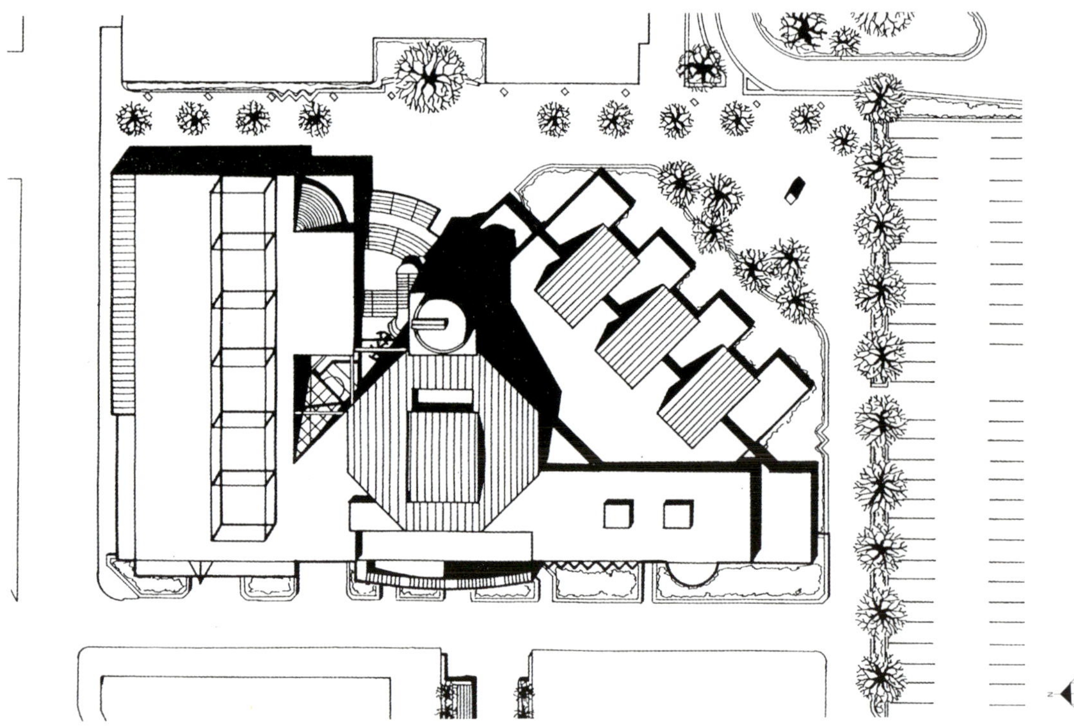

图书馆内部

図書館内部

Inside the library

科学馆内景
科学館内部
Inside the science museum

科学馆内部

科学館内部

Inside the science museum

科学馆的通路

科学館へのアプローチ

Approach to the science museum

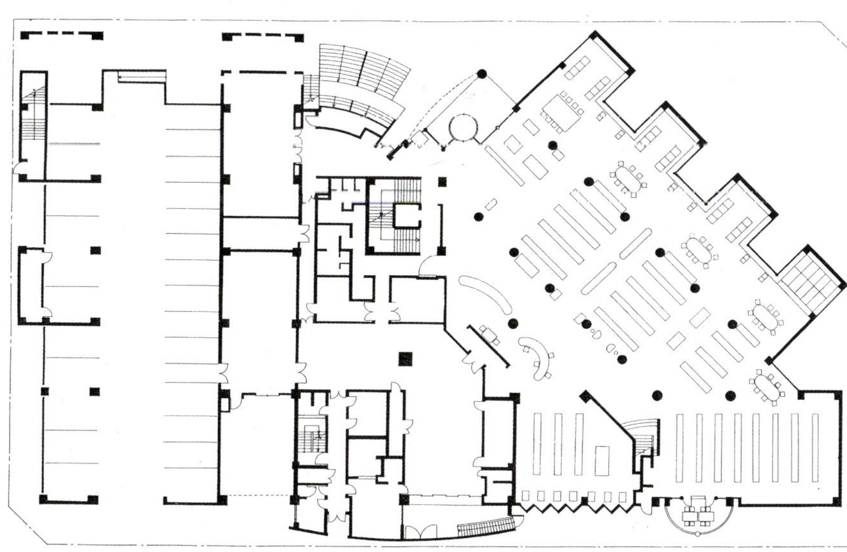

秋田市太平山自然学习中心

所 在 地　秋田县秋田市
竣工时间　2003年8月
建筑结构　钢筋混凝土结构,部分为钢结构、木结构 / 地上3层，地下1层
用地面积　42209.77m^2
占地面积　4098.71m^2
建筑面积　5284.51m^2

本设施位于优美的自然环境之中，为青少年的健康成长以及市民的终生学习提供了与自然接触以及野外活动和创作活动的场所。全体设施通过联络渡桥连接成为一个整体，形成了如同在森林当中一样的开放的空间结构。

秋田市太平山自然学習センター

所 在 地　秋田県秋田市
竣 工 年　2003年8月
規　　模　RC造一部S造、W造／地上3階 地下1階
敷地面積　42,209.77m^2
建築面積　4,098.71m^2
延床面積　5,284.51m^2

豊かな自然環境の中で、自然とのふれあいや野外活動、創作活動を通じ、青少年の健全育成と市民の生涯学習の施設である。施設は連絡ブリッジで一体的に結ばれており、森の中にいるような開放的空間構成を持っている。

Akita Taiheizan Nature Learning Center

Location　Akita City, Akita Prefecture
Completion　August 2003
Structure　RC, part Steel, W/1 floor below-grade, 3 above-ground floors
Site area　42,209.77m^2
Building area　4,098.71m^2
Total floor area　5,284.51m^2

This is a facility for youth people's healthy growth and citizen's life-long learning through contact with nature, outdoor activities and artistic works in the rich natural environments. The whole facility is united by a connection corridor, therefore, visitors can feel like being in a woodland because of openness it has.

主入口
メインエントランス
Main entrance

木桥

木橋
Wooden bridge

眺望大屋面活动楼和本馆

大屋根研修棟と本館を望む
Viewing sheltered meeting area and main building
akita large

活动室
研修スペース
Activity space

卧室
宿泊室
Bedroom

工作室
ワークショップ
Workshop room

屋面广场的内部

大屋根広場の内部

Inside the sheltered meeting area

兵库县南但马自然学校

所 在 地	兵库县朝来郡
竣工时间	1994年5月
建筑结构	钢筋混凝土结构，部分为木结构
用地面积	$1200000m^2$
占地面积	$6429.22m^2$

兵库县正在整备能够为本县所属小学校的全体5年级学生提供可以进行五宿六天的自然学习体验的自然学校。本设施就是作为其模式而进行设计的。孩子们住宿的建筑楼也不同于以往意义上的"住宿楼"，而是采取了活动的"生活楼"的形式，设计成具有宽敞的吹拔空间的木结构建筑。

兵庫県南但馬自然学校

所 在 地	兵庫県朝来郡
竣 工 年	1994年5月
規　　模	RC造木造
敷地面積	$1,200,000m^2$
延床面積	$6,429.22m^2$

兵庫県は県下の小学校5年生全員に、5泊6日の自然学習体験のできる自然学校の整備を進めている。本施設はそのモデルとしてつくられた。こども達の宿泊する棟は従来の宿泊するための"宿泊棟"という形式ではなく、活動的な"生活棟"という形式を取り、広い吹き抜け空間をもった木造の建物とした。

Minami Tajima Nature School

Location	Asago-gun, Hyogo Prefecture
Completion	May 1994
Structure	RC, W
Site area	$1,200,000m^2$
Total floor area	$6,429.22m^2$

Hyogo Prefecture is promoting facilities in non-urban areas where all fifth grade children from elementary schools in the prefecture can stay for up to five nights, in order to learn from nature what they cannot gain in the classroom, and this was built as a model facility. Unlike conventional dormitory-like buildings for children, a residential form of building was adopted for the main building, where the children can learn while engaged in various activities, and this timber building was given a large open well.

暮色中的大屋顶广场

夕暮れの大屋根広場

The sheltered meeting area at dusk

生活楼

生活棟
Residential building.

大屋顶广场内部

大屋根広場内部

Inside the sheltered meeting area

佐久市儿童未来馆

所 在 地	长野县佐久市
竣工时间	2001 年 2 月
建筑结构	钢筋混凝土结构，钢结构 / 地上 3 层，地下 1 层
用地面积	7112m²
占地面积	3507m²
建筑面积	2117m²

本建筑在外观形式上采用圆盘的形状，象征着孩子们的未来。大小两座拱顶建筑分别是展示馆和天象仪馆。建筑的内部空间是以“创造未来”——进化的宇宙、地球、生命为主题，通过围绕着 3 层结构的不规则塔而展开的。

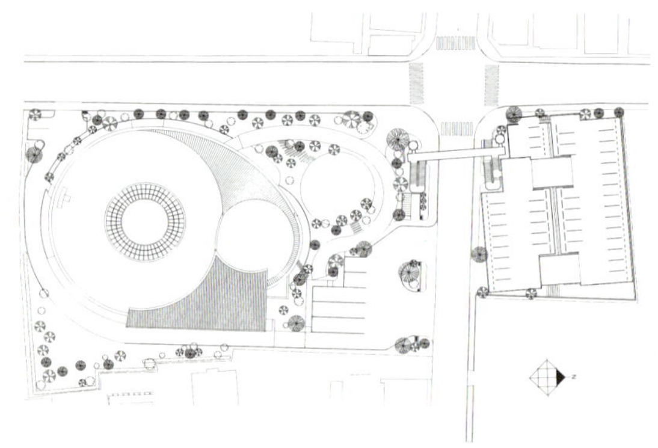

佐久市子ども未来館

所 在 地	長野県佐久市
竣 工 年	2001 年 2 月
規　　模	RC 造、S 造／地上 3 階、地下 1 階
敷地面積	7,112m²
建築面積	3,507m²
延床面積	2,117m²

外観は、こども達の未来を象徴して、円盤のような形状とした。大小 2 つのドームは、展示棟とプラネタリウムである。建物内部は「未来への創造」－進化する宇宙・地球・生命－をテーマに、3 層構造のフラクタルタワーを巡るように展示展開している。

Saku Children's Science Dome for the Future

Location	Saku City, Nagano Prefecture
Completion	February 2001
Structure	RC, Steel/1 floor below-grade, 3 above-ground floors
Site area	7,112m²
Building area	3,507m²
Total floor area	2,117m²

Symbolizing the future of children, the outer appearance was constructed in a disk-like shape. There are two domes, large and small, one for the exhibition arena and one for the planetarium. Exhibitions are arranged to revolve around the three layer "fractal tower" inside the building, on the theme of "creation for the future", furthering the universe, earth and life.

中央是展示室“不规则形塔”

中央は展示室「フラクタルタワー」

Exhibition room, "Fractal Tower", in the center

外观。由天文馆和展示楼的两座圆拱顶所组成

外観.プラネタリウムと展示棟の二つの丸いドームから成る

View of the external building. It consists of two domes: a planetarium and science museum block

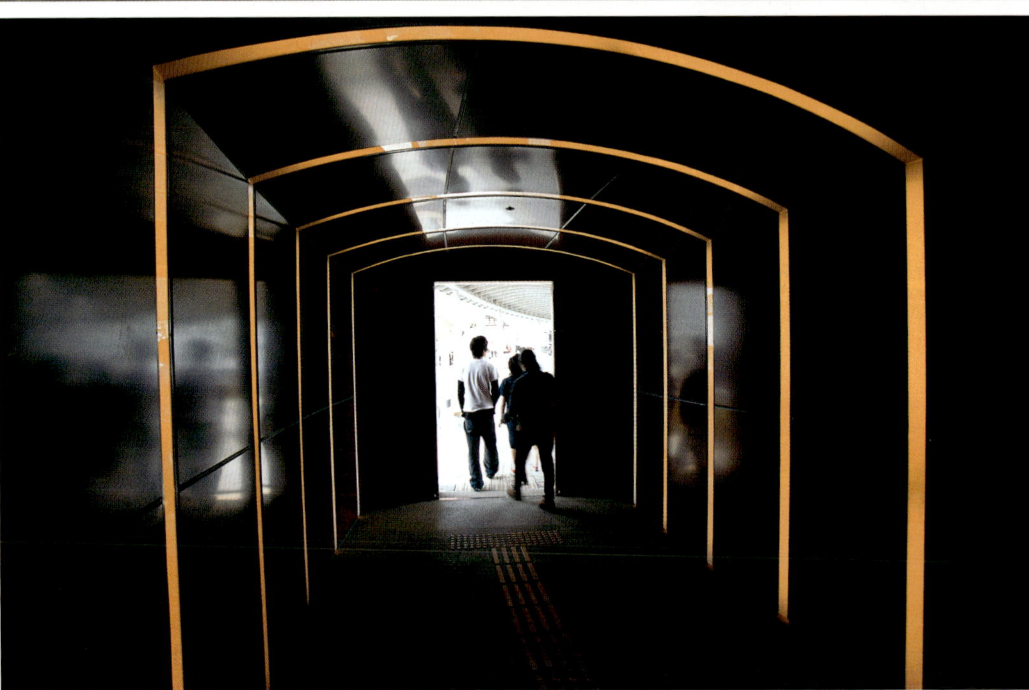

屋外塔状游具
屋外塔状遊具
Outdoor tower play structure

连接停车场和本馆的步道桥和游戏玩具

駐車場と本館をつなぐ歩道橋と遊具

Pedestrian bridge connecting the car park and main building, and play structure

滨松儿童馆

所 在 地	静冈县滨松市
竣工时间	2001 年 11 月
建筑结构	钢筋混凝土结构，部分为钢结构/地上15层建筑当中的6、7层部分
用地面积	4673m²
占地面积	2315.5m²(6层) 689.47m²(7层)

为了搞活中心城市街区，滨松市对于大型商业设施进行了再开发，在底层部分与高层住宅部分之间的6、7层设置了城市型的儿童及育儿支援设施。在中央部分的吹拔空间设置了从底层的商业设施就能看到的象征性标志物，并被全长85m的空中隧道游戏玩具所围绕。同时也设置了大型的网状游戏玩具——宇宙塔和儿童电梯等移动游戏玩具，6、7层通过游戏动线连接成为一体。

浜松こども館

所 在 地	静岡県浜松市
竣 工 年	2001 年 11 月
規　　模	RC 造一部 S 造／地上 15 階のうち 6・7 階部分
敷地面積	4,673m²
延床面積	2,315.5m²（6 階）689.47m²（7 階）

浜松市が、中心市街地の活性化に向けた大型商業施設の再開発事業に際して、低層部の商業施設と高層部の住居棟の間、6～7階に設けた都市型の児童施設・子育て支援施設である。中央の吹き抜けには、下の商業施設からも見上げられるシンボルモニュメントを配置し、これらを全長85mのあそび機能を持った空中トンネルが取り囲む。大型ネット遊具·コスモタワーやこどもエレベーターなど活動的な遊具を設け、6階と7階をあそびの動線で連続させている。

Hamamatsu Children's Center

Location	Hamamatsu City, Shizuoka Prefecture
Completion	November 2001
Structure	RC, part Steel/6-7th floor in 15 above-ground floors
Site area	4,673m²
Total floor area	2,315.5m²（6th floor） 689.47m²（7th floor）

An urban type children's facility and childcare support facility established on the 6th and 7th floors between lower level commercial facilities and upper level residential sections in a Hamamatsu City redevelopment project for large-sized commercial facilities that aims at vitalizing the city center area. Symbolic monuments are arranged in the central atrium, visible from below in the commercial facilities, and these are surrounded by air tunnels for play totaling in length 85 m. Active play equipment such as jumbo net play equipment, a cosmos tower and a children's elevator are arranged and the 6th and 7th floors are serialized with a play flow line.

屋顶游戏玩具：三只眼小僧
屋上遊具：三ツ目こぞう
Rooftop play structure, "Kid with Three Eyes"

施入口

設入口

ce of the com-
building

连接6、7层的巨大网状游戏玩具

2. 6階、7階をつなぐ巨大なネット遊具

Huge net play structure connecting 6th and 7th floor

屋顶广场
屋上広場
Roof-top play area

儿童升降机，上下连接 6、7 层的大型游戏玩具
こどもエレベーター .6 階、7 階を上下につなぐ大型遊具
Kids' elevator connecting 6th and 7th floor

儿童馆入口
こども館入口
Entrance of the children's center

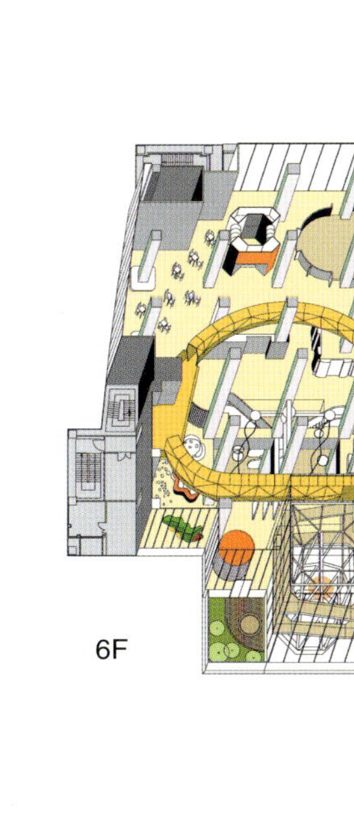

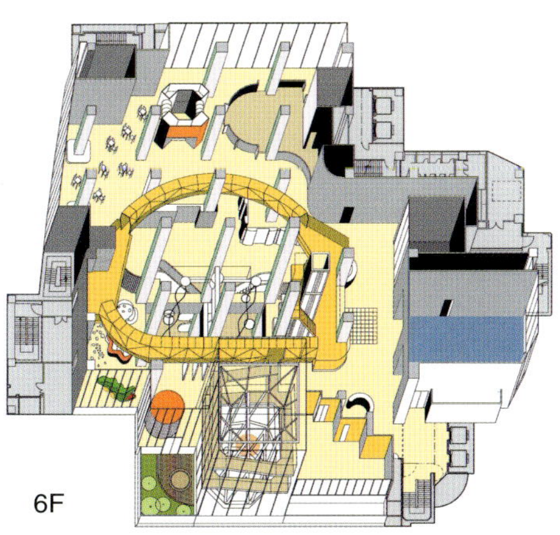

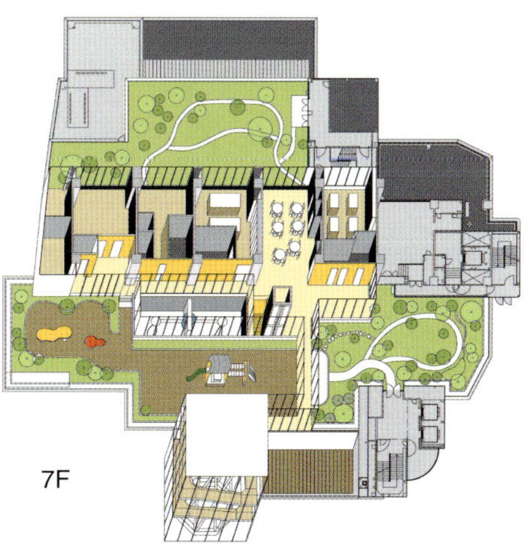

面向富士山的单坡屋顶外观

野中ザウルス.富士山に向かって上る片流れの屋根が連続する外観

The Nonaka Zaurus. The facade of the roof connects with Mt. Fuji

野中丸，连廊也可以成为画廊

野中丸.手前のデッキはギャラリーにもなる

The Nonaka Maru. The front deck can be used as a gallery

野中保育园

所 在 地	静冈县富士宫市
竣工时间	1972年8月（野中沃尔斯）
建筑构造	钢结构/地上2（野中沃尔斯）
用地面积	6600m²
占地面积	592m²
建筑面积	647m²

该保育园地处能够眺望到富士山的小山丘上。在设计中遵循以大地保育的理念为设计宗旨，并将其进行建筑化。连接位于建筑北侧的被称为散步空间的回廊和保育室的是2层的散步路，通过该散步路形成室内回游的移动路线。

新园舍（野中丸）

所 在 地	静冈县富士宫市
竣工时间	1981年3月（野中丸）
建筑构造	钢结构/地上1层（野中丸）
用地面积	4485m²（野中丸）
占地面积	328m²（野中丸）
总建筑面积	325m²（野中丸）

该建筑是野中保育园（野中活尔斯）的扩建工程。保育室之间的隔墙是可移动的，因此也可以作为上型的体育馆和剧场来使用。

野中保育園 「野中ザウルス」

所 在 地	静岡県富士宮市
竣 工 年	1972年8月「野中ザウルス」
規　　模	S造／地上2階「野中ザウルス」
敷地面積	6,600m²「野中ザウルス」
建築面積	592m²「野中ザウルス」
延床面積	647m²「野中ザウルス」

富士山を望む丘の上に立つ保育園。大地保育という保育理念の建築化をめざした。北側にプロムナードと称する内廊と保育室をつなぐ2階のプロムナードによって回遊動線を形成している。

新園舎「野中丸」

所 在 地	静岡県富士宮市
竣 工 年	1981年3月「野中丸」
規　　模	S造／地上1階「野中丸」
敷地面積	4,485m²「野中丸」
建築面積	328m²「野中丸」
延床面積	325m²「野中丸」

野中保育園（野中ザウルス）の増築。保育室の隔壁が可動で、小さな体育館や劇場にも変わる。

Nonaka Day Nursery 'Nonaka Zaurus'

Location	Fujinomiya City, Shizuoka Prefecture
Completion	August 1972
Structure	Steel/2 above-ground floors
Site area	6,600m²;
Building area	592m²
Total floor area	647m²

A preschool standing on a hill in view of Mt. Fuji that gives concrete form to the ideal of a nursery, that is, a nursery on the land. A flow line circulation is formed by a promenade on the north side connecting the inner corridor and the nursery rooms.

Nonaka Day Nursery 'Nonaka Maru'

Location	Fujinomiya City, Shizuoka Prefecture
Completion	March 1981
Structure	Steel/1 above-ground floor
Site area	4,485m²
Building area	328m²
Total floor area	325m²

Expansion of Nonaka Zaurus. Partition walls in the nursery room are movable, so it can be transformed into a small gym or theater.

散步路的平均幅度为 3m
野中ザウルス.プロムナードの平均幅は3m
The Nonaka Zaurus. The average width of the promenade is 3m

野中丸，如同仓库一样的绿色箱体的外观
野中丸 倉庫のような緑の箱の外観
The Nonaka Maru. The facade of green box like a warehouse

野中沃尔斯内部
野中ザウルス内部
Inside the Nonaka Zaurus

野中丸，舞台成为孩子们表现自己的场所
野中丸.ステージがこどもたちの表現の場となる
The Nonaka Maru. The stage becomes a space of children's performance

野餐

野中ザウルス.外での食事
The Nonaka Zaurus. Having lunch outside

野中沃尔斯室内

野中ザウルス内観
Inside the Nonaka Zaurus

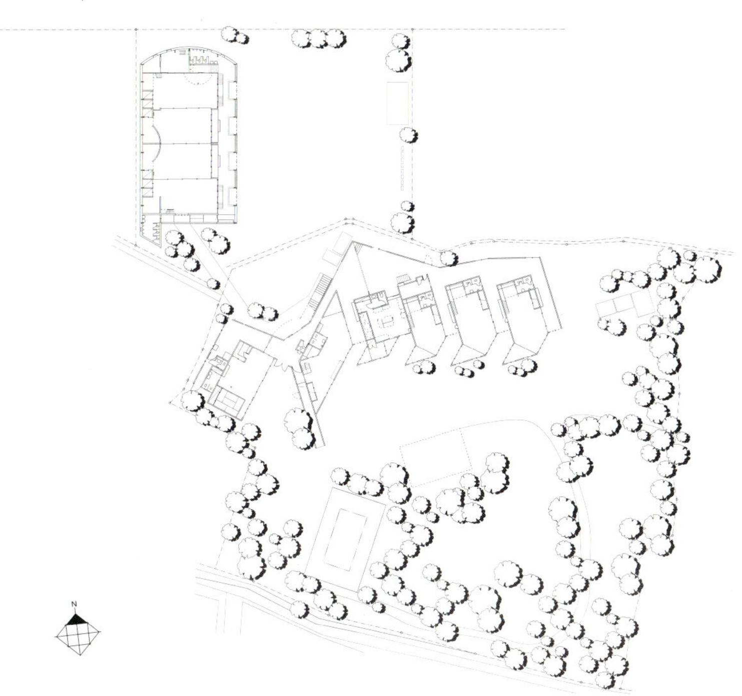

红色渡桥从大屋顶下的广场穿过

大屋根の下の広場には赤いブリッジが走る

There is a red bridge running in the plaza under the big roof

石川少年自然之家

所 在 地	冲绳县石川市
竣工时间	1975 年 3 月
建筑结构	钢筋混凝土结构 / 地上 3 层
用地面积	$121361m^2$
占地面积	$1197m^2$
建筑面积	$2172m^2$

该建筑是建在石川山麓之上的，能够容纳 200 名儿童进行体验自然的住宿设施，通过以大屋顶为中心被分隔成 4 部分的住宿楼而组成的有气孔的建筑。

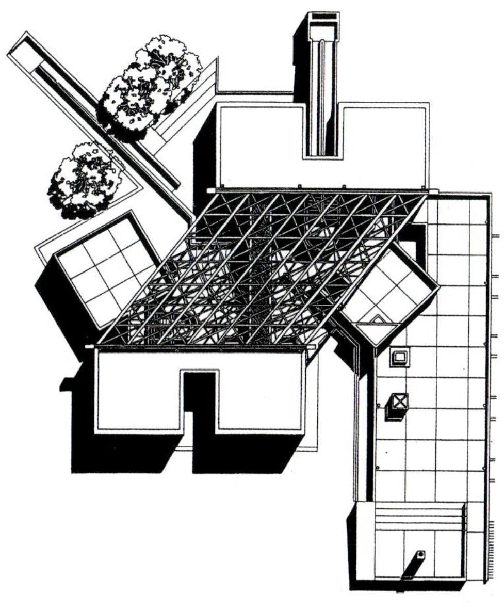

石川少年自然の家

所在地	沖縄県石川市
竣工年	1975年3月
規　模	RC造／地上3階
敷地面積	121,361m²
建築面積	1,197m²
延床面積	2,172m²

石川岳の麓につくられた、収容200人のこどものための自然体験宿泊施設。大屋根を中心に4つに分割された宿泊棟により構成されたポーラスな建築である。

Ishikawa Children's Nature School

Location	Ishikawa City, Okinawa Prefecture
Completion	March 1975
Structure	RC/3 above-ground floors
Site area	121,361m²
Building area	1,197m²
Total floor area	2,172m²

This nature experience lodging facility for children has a capacity of 200 and is constructed at the foot of Mt. Ishikawa. The lodging section, divided into four and centered under a massive roof is of porous type construction.

南侧外观，因为上部的出挑屋顶部分是游泳池所以向上翻起

南側外観.上部の張り出しは屋上にプールがあるため折り返しをつけている

Southern exterior. The overhang at the top is folded back along the edge of the rooftop pool

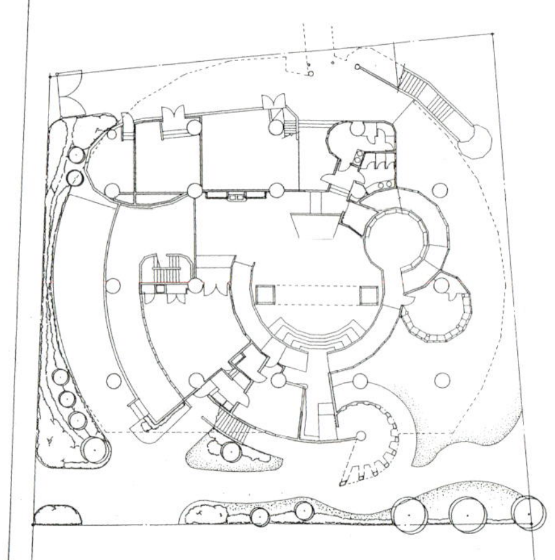

相模原市星之丘儿童中心

所 在 地	神奈川县相模原市
竣工时间	1993 年 7 月
建筑结构	钢框钢筋混凝土结构/地上3层
用地面积	1629.11m²
占地面积	706.51m²

为了实现在市内53所小学校的每一所小学校内都要设置一所儿童中心的计划，本设施特意按照其模式进行设计。儿童中心预计每天接待200名来访者。在平面布局上以位于中央部分的游戏室为中心，并以环绕着它的形式设置了图书区域等各种内容的活动区域。游戏室设置玩具的桥，使整个建筑都成为一个大的具有环游结构的游戏装置。

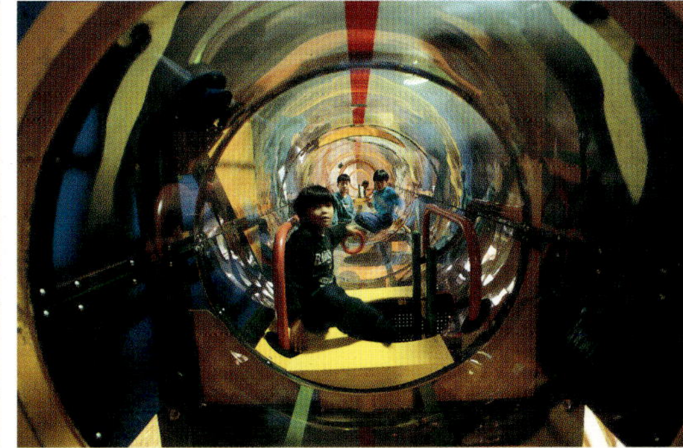

相模原市星ヶ丘こどもセンター

所 在 地	神奈川県相模原市
竣 工 年	1993年7月
規　　模	SRC造／地上3階
敷地面積	1,629.11m²
延床面積	706.51m²

市内の53校の小学校に1校ずつ児童センターを作る計画があり、この施設はそのモデルとして建設された。児童センターは1日約200人のこども達が来館する。中央にある遊戯室を中心として、それにとりつく形で図書コーナーをはじめ様々なコーナーが作られている。遊具の橋が遊戯室の空間をはしり、遊環構造建築となっている。

Sagamihara Municipal Hoshigaoka Children's Center

Location	Sagamihara City, Kanagawa Prefecture
Completion	July 1993
Structure	SRC/3 above-ground floors
Site area	1,629.11m²
Total floor area	706.51m²

Individual children's centers are planned for the city's jurisdiction, and this particular facility was built as a model for these. Approximately two hundred children visit the center in one day. It has a playroom at its center with various other activity areas including a library wrapping around it. A bridge with play features runs through the playroom and it is based on circular play system.

屋顶泳池

屋上プール

Roof-top swimming pool

建筑内部就是一个整体的环游结构

内部は全体として遊環構造をもった建築となっている

Overall the architecture has Circular Play System

渡桥的正面，下部是儿童剧场

ブリッジを正面より見る.下部はこども劇場

View of the bridge on the front. There is a children's theater under the bridge

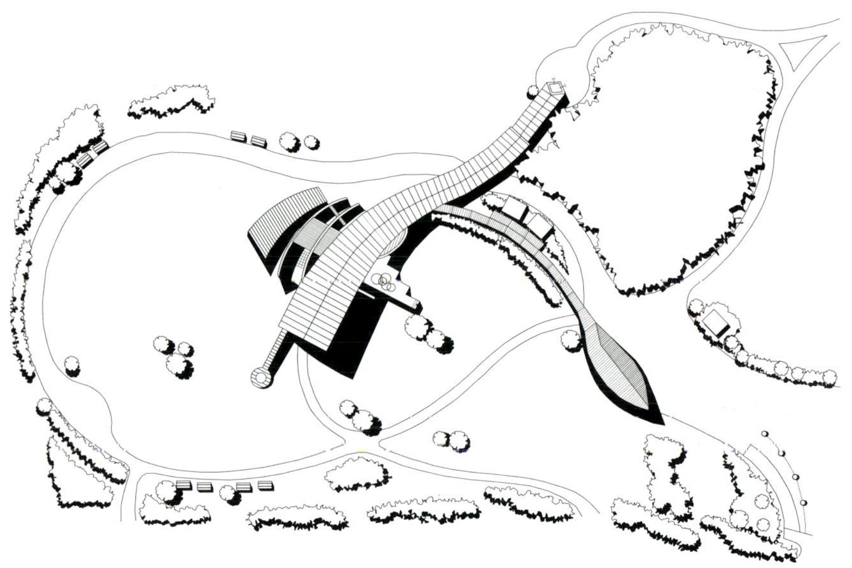

旭川春光台（风之子馆）

所 在 地	北海道旭川市
竣工时间	1994 年 7 月
建筑结构	木结构 · 钢结构
用地面积	约 8000m^2
占地面积	485m^2

该建筑位于旭川市的新住宅开发地——春光台，是日本第一座带有屋顶的木制双重结构的游戏设施。作为拥有漫长冬季的雪国的游戏设施，该建筑拥有巨大的屋顶，并且开发了尽可能地长时间能够被利用的游戏玩具。

渡桥内部，无秩序的游戏空间
ブリッジ内部.アナーキーなあそびの空間
Inside the bridge. It is an anarchy play space

旭川春光台「風の子館」

所在地	北海道旭川市
竣工年	1994年7月
規　模	木造・S造
敷地面積	約 8,000m²
延床面積	485m²

旭川市の新しい住宅開発地・春光台にある日本ではじめて構想された木製二重構造の屋根付き遊具である。冬の長い雪国の遊具として大きな屋根を持ち、出来るだけ長い季節あそべる遊具として開発された。

Asahikawa Shunkodai Park 'Kazenoko-Kan'

Location	Asahikawa City, Hokkaido
Completion	July 1994
Structure	W, Steel
Site area	Approx. 8,000m²
Total floor area	485m²

This is the first two story roofed wooden play structure designed in Japan, located in the new residential development of Shunkodai in Asahikawa. It was developed as a play structure with a large roof to make it usable as late as possible in the season in an area known for its heavy winter snowfall.

可以进行各种游戏的广场
さまざまなあそびができる広場
A square where children play various things

从操场看校舍
グラウンドから校舎を見る
School building seen from the ground

从校舍看操场的景观
校舎側から見たグラウンドの風景
The ground seen from the building

川崎市向丘小学校

所 在 地	川崎市宫前区
竣工时间	1997 年 3 月
建筑结构	钢筋混凝土结构，部分为钢框架钢筋混凝土结构 / 地上 4 层，屋顶间 1 层
用地面积	9646m^2
占地面积	3441m^2
建筑面积	8352m^2

在建筑内设计了能够环绕一周的走廊，即使是在外廊空间也没有阻挡，因此在这里孩子们的行动不会受到任何限制。在教学楼与体育馆之间是被称为伽勒利阿的半室外广场。本设计的目的就是力图为孩子们创造一个不仅可以感受到室内外空间的各种变化，同时也可以对身边的同学、老师以及每一天都在发生变化的自然能够有重新认识和了解的空间。

川崎市立向丘小学校

所在地	川崎市宮前区
竣工年	1997年3月
規　模	RC造・一部SRC造／地上4階・塔屋1階
敷地面積	9,646m²
建築面積	3,441m²
延床面積	8,352m²

建物内をぐるりと一周できるような廊下、外廊の空間にも行き止まりを作らず、こども達の行動を制限しない動線計画、教室棟と体育館の間のガレリアと呼ぶ半野外広場等、こども達が内外の様々な変化を体感し、友達や先生、日々変化する自然との新たな出会いを発見できる空間として意図されている。

Mukaigaoka Elementary School, Kawasaki City

Location	Miyamae-ku, Kawasaki City
Completion	March 1997
Structure	RC, part SRC/4 above-ground floors
Site area	9,646m²
Building area	3,441m²
Total floor area	8,352m²

Inside the building, the corridor allows one to make a complete circle with no dead ends, a plan that eliminates boundaries on children's movement. The indoor-outdoor plaza between the school building and the gymnasium, known as the "Galleria", permits children to experience various indoor / outdoor changes-a space to encourage children to discover new things in contact with their friends, teacher, and the daily changes created by nature.

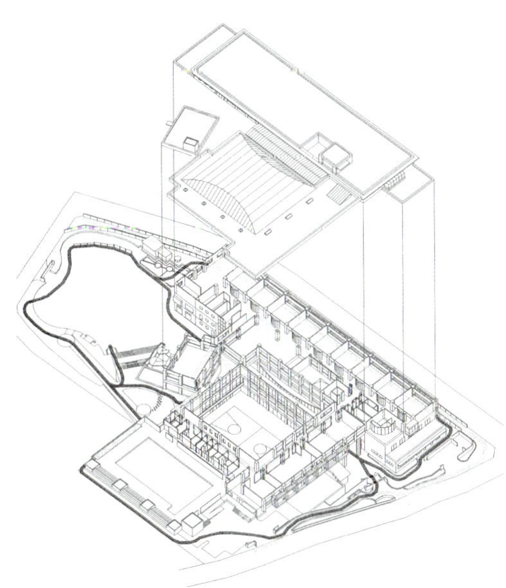

休息空间，有小型公园的通学道路
憩いの空間，ポケットパークのある通学路
A space for resting, a street with a pocket park

建筑外观
建物外観
External view of the building

爱媛儿童城

所 在 地　　爱媛县松山市
竣工时间　　1998 年 7 月
建筑结构　　钢筋混凝土结构 / 地上 2 层
用地面积　　350000m²
占地面积　　3019.81m²
建筑面积　　4457m²

在建筑的空间设计上充分利用地形的高低差约18m 的倾斜地势，大致设定了三个标高，整体形态呈台阶状。入口 · 剧场 · 工作室等不同标高层上的空间分别通过坡道和游戏玩具连接起来，形成室内空间顺畅的连续性。

屋顶游戏玩具
屋上遊具
Rooftop play structure

1层的多媒体区域
1階のマルチメディアコーナー
Multi-media room on the first floor

えひめこどもの城

所在地	愛媛県松山市
竣工年	1998年7月
規　模	RC造／地上2階
敷地面積	350,000m²
建築面積	3,019.81m²
延床面積	4,457m²

高低差約18mの傾斜地を利用して、大きく三つのレベルが設定され、全体としては段上の構成である。エントランスレベル・シアターレベル・アトリエレベルのそれぞれをスロープと遊具によって連続させ、室内の展開を演出している。

Ehime Children's Castle

Location	Matsuyama City, Ehime Prefecture
Completion	July 1998
Structure	RC/2 above-ground floors
Site area	350,000m²
Building area	3,019.81m²
Total floor area	4,457m²

Three main levels are established in vertical intervals of about 18m using the sloping ground. Overall construction is above the stage. The entrance, theater and atelier levels all unfold consecutively with slopes and play equipment, and exhibitions are staged inside the rooms.

多媒体区域上部是绳网游戏玩具
マルチメディアコーナー上部はネット遊具になっている
There is a net feature above the multi-media room

建筑前面的太阳钟广场
建物正面の日時計広場
Sun dial square in front of the building

通过电梯进入屋顶层
エレベーターで屋上階へ
Going to the top floor by elevator

福井县儿童科学馆

所 在 地	福井县坂井郡
竣工时间	1998 年 7 月
建筑结构	钢筋混凝土结构，钢结构，木结构 / 地上 3 层
用地面积	55826m^2
占地面积	5615m^2
建筑面积	6874m^2
获奖经历	2000 年商业环境设计奖， 入选 2001 年建筑学会作品选集

本儿童科学馆位于福井县中央部的福井市北部住宅区春江町内。其中的游戏区域以被称为恐龙广场的空间为主，设置了连环画区域和婴儿活动区域。北侧部分为机械室，在它的中央是展示化学实验的区域，二层部分是以化学实验为中心。南侧面向草坪广场，在空间设计上力图使建筑室内外空间融合成为一个整体。

福井県児童科学館

所在地	福井県坂井郡
竣工年月	1998年7月
規模	SRC、S造、木造／地上3階
敷地面積	55,826m²
建築面積	5,615m²
延床面積	6,874m²
受賞歴	2000.商環境デザイン賞 2001.日本建築学会作品選集

福井県の中央部福井市の北に位置するベッドタウン春江町に建設された。あそびゾーンは恐竜広場と呼ばれるゴム底をメインとし、絵本のコーナー、乳児コーナーなども設けられている。北側部分は機械室で、その中間に化学実験ショーのコーナー、2階部分は化学実験が中心。南側は芝生広場に面し、外部空間と一体となるよう計画されている。

Fukui Children's Science Center

Location	Sakai-gun, Fukui Prefecture
Completion	July 1998
Structure	SRC, Steel, W/3 above-ground floors
Site area	55,826m²
Building area	5,615m²
Total floor area	6,874m²
Awards	2000 JCD Award; 2001 Architectural Institute of Japan Selected Architectural Designs

Constructed in the city of Harue, a bed town north of Fukui City, the center of Fukui Prefecture, the center includes as its main part a rubber-surfaced play zone called "Dinosaur Square", along with a picture book corner and infants corner, etc. On the northern side is a machinery room, and between these areas is a corner for chemistry experiment demonstrations. The second floor is mainly for chemical experiments. The southern side faces a lawn square and is designed as a single unit with the external space.

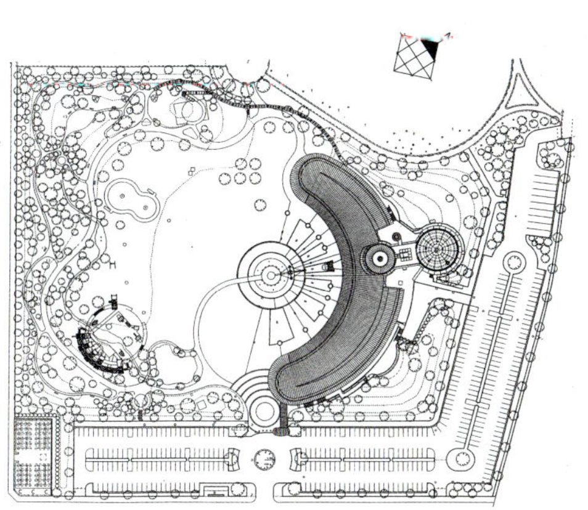

室外游戏玩具，孩子们的白云
屋外遊具「こどもの雲」
Rooftop play structure "Kids' Clouds"

入口
エントランス
Entrance.

中国北京市科普乐园

所 在 地　中国北京市
竣工时间　1991 年 5 月
建筑结构　砖混结构，部分使用钢材，预制板
用地面积　7000m²

该建筑用地紧邻北京宋庆龄儿童科学馆。在用地面积约1000m²的范围内设计了8个科学游戏玩具，并采用在北京具有代表性的砖作为建筑材料，而施工则是由中国方面进行的。在设计手法上采用错觉之门等各种各样视觉效果的变化为主题，力图使人们在游戏中体验科学原理以及科学的趣味。图为科学游戏玩具的代表实例。

中国北京市科普楽園

所 在 地　中国北京市
竣工年月　1991 年 5 月
規　　模　レンガ一部 S 造、FRP
敷地面積　7,000m²

北京宋慶齢児童科学館に隣接した敷地約1,000平方メートルに８つの科学的な遊具が計画された。錯覚のゲート等、様々な視覚的な変化をテーマとし、あそびながら、科学の原理やおもしろさが体験できるように意図されている。科学遊具の代表例である。

Beijing Song Qingling Children's Science Park

Location　Beijing, China
Completion　May 1991
Structure　Brick, part Steel, FRP
Site area　7,000m²

Eight pieces of play equipment to help explain scientific effects and principles were planned for this 1000 sqm site adjusting Beijing Song-Qing-Ling Children's Science Museum, with the idea of getting children to actually experience some of the principals of science while they play based on a theme of various visual changes such as a gate of hallucination. It is therefore representative of a piece of scientifically orientated play equipment.。。

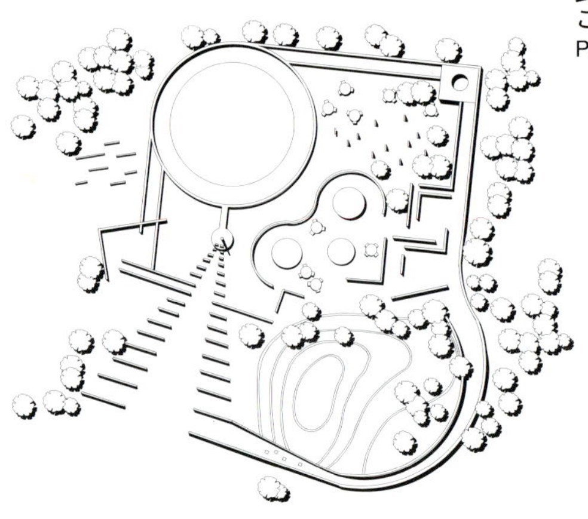

由带孔的墙壁所构成的环游结构
穴のある壁による遊環構造の構成
Design of porous wall is based on 'circular play system'

此项目摄影：环境设计研究所
この頁の撮影：環境デザイン研究所
Photo: Environment Design Institute

在圆环形状的游戏玩具内部暗藏着各种各样的机关
円環状の平面形をもつ遊具のなかには、いろいろな仕掛けが隠されている
Various gadgets are hidden inside this circular play structure

开放式天文馆
オープンプラネタリウム
Open planetarium

回音管道
エコーチューブ
Echo tube

摄影：寿
撮影：コトブキ
Photo: Kotobuki

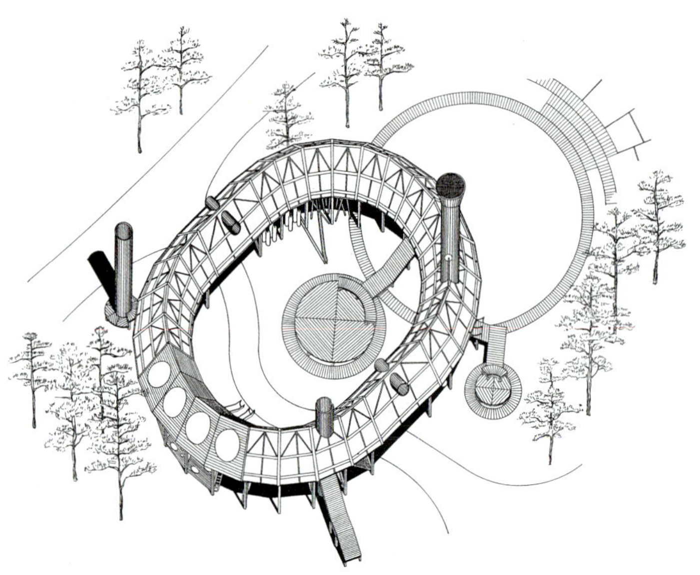

信州博登山广场圆环游戏装置

所 在 地	长野县松本市
竣工时间	1993 年 7 月
建筑结构	木结构，部分为钢结构
建筑面积	1250m^2
获奖经历	1993 年日本万国博览大奖

该圆环游戏玩具是一座周长80m的建筑性质的游戏玩具。该游戏玩具是以探索自然为主题而建造的，它不仅可以使人们在游戏的同时能够听到风声和鸟鸣，而且还可以通过开敞的天象仪观察到松本四季的星空夜景。在博览会之后，该游戏玩具将作为永久设施而被保留下来。

信州博アルピコ広場円環遊具

所 在 地	長野県松本市
竣 工 年	1993 年 7 月
規　 模	木造一部 S 造
敷地面積	1,250m^2
受 賞 歴	1993.ジャパンエキスポ大賞

一周80mの円環状の建築的な遊具であり、あそびながら風の音、鳥の声を聞き、松本の四季の星空をオープンプラネタリウムで見たりというように、自然発見遊具として構想された。博覧会後、恒久施設として残された。

Shinshu Expo Alpico Play Structure

Location	Matsumoto City, Nagano Prefecture
Completion	July 1993
Structure	W, part Steel
Site area	1,250m^2
Awards	1993 Japan Expo. Grand Prix

This circular piece of architectural play equipment is 80 meters in length and was planned as nature discovery play equipment to help children to discover some of the sound of the wind, birds singing and to be able to recognize the changing seasons in and around Matsumoto both during the day and at night. After the expo closed, it was retained as a permanent feature.

安装在游戏玩具屋顶上的四季的天象仪
遊具屋根に取り付けられた四季のプラネタリウム
Star maps of the four seasons set amongst the rafters of this play structure

通过3连传话管进行交谈
3連の伝声管で会話
Conversation on the speaking tube

从风之道眺望风之家
風の道より風の家を望む
View of "Road of Wind" from "House of Wind"

姬路御立公园龙卷风大道

所 在 地	兵库县姬路市御立西
竣工时间	1993 年 3 月
建筑结构	钢柱，木结构
用地面积	6300m²
获奖经历	1994 年日本造园顾问协会奖

御立公园是位于姬路市郊外的交通公园，该设施就建于御立公园内，属于巨型的游戏玩具。在广场上是具有冲击效果的龙卷风的动感造型，它为孩子们的玩耍起到引爆剂的作用。

姫路御立公園たつまきロード

所 在 地	兵庫県姫路市御立西
竣 工 年	1993 年 3 月
規　 模	スチール製柱、木造
敷地面積	6,300m²
受 賞 歴	1994.日本造園コンサルタント協会賞

姫路市の郊外交通公園として整備されていた御立公園に設置された巨大遊具型の遊具である。広場にたつまきが起こったような動的で、衝撃的な造形である。こども達の遊びの起爆剤となるように動線がデザインされている。

Himeji Mitate Park Tornado Road

Location	Mitatenishi Himeji City, Hyogo Prefecture
Completion	March 1993
Structure	Steel, W
Site area	6,300m²
Awards	1994 CLA Award

This large scale decked play apparatus has been set up in Mitate Park which has been completed as a "transport park" in the suburbs of Himeji. The form has an impact as if a tornado moves around the filed. The course of the decking itself is designed to trigger and stimulate all kinds of play.

设置在空之道上的迂回道路
空の道に設置されたバイパス通路
By-pass in "Road of Sky"

宝德幼儿园

所 在 地	福岛县磐城市
竣工时间	2000年4月（儿童森林保育中心）2002年6月（园舍）
建筑结构	木结构/地上2层（儿童森林保育中心）钢筋混凝土结构，部分为钢结构/地上2层（园舍）
用地面积	3057m^2（儿童森林保育中心） 7159.65m^2（园舍）
占地面积	102.89m^2（儿童森林保育中心） 713.08m^2（园舍）
建筑面积	159m^2（儿童森林保育中心） 1269.64m^2（园舍）
获奖经历	2003年设计参与/SCN奖

该幼儿园是由寺院运营的，以〝与家人在一起〞这个理念为开办宗旨。为了实现这个宗旨，对相当于〝村庄〞概念的新园舍（2002年进行改建）、保育中心（2002年新建）以及相当于〝村庄中的广场〞的庭院（2002年开始建造）进行了设计，提出了〝成长农场〞的设计理念。

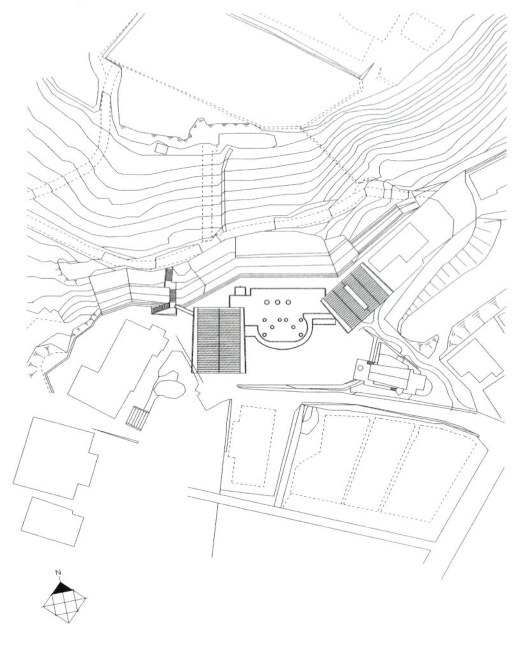

建筑室内，上部布满了网状的游戏玩具

園舎内部.上部にはネット遊具がはりめぐらされる
Inside the building. There is a net play feature above the corridor

ほうとく幼稚園

所在地	福島県いわき市
竣工年	2000年4月（子供の森保育センター）2002年6月（園舎）
規　模	木造／地上2階（子供の森保育センター）RC造一部S造／地上2階（園舎）
敷地面積	3,057m²（子供の森保育センター）7,159.65m²（園舎）
建築面積	102.89m²（子供の森保育センター）713.08m²（園舎）
延床面積	159m²（子供の森保育センター）1,269.64m²（園舎）
受賞歴	2003.DesignShare/SCN 賞

この幼稚園はお寺が運営しており「家族と一緒」という理念を掲げている。その理念に配慮して「村」にあたる新園舎（2002年に改築）、保育センター（2000年に新築）、そして「村の広場」となる園庭（2002年〜）の設計を行い、「成長するファーム」を提案している。

Houtoku Kindergarten

Location	Iwaki City, Fukushima Prefecture
Completion	April 2000 (Nursery Center); June 2002 (School Building)
Structure	W/2 above-ground floors (Nursery Center); RC, part Steel/2 above-ground floors (School Building)
Site area	3,057m² (Nursery Center); 7,159.65m² (School Building)
Building area	102.89m² (Nursery Center); 713.08m² (School Building)
Total floor area	159m² (Nursery Center); 1,269.64m² (School Building)
Awards	2003 DesignShare/SCN Award

The kindergarten has been run by a Buddhism temple with policy of 'family together'. While considering their policy, we designed the new school building (built in 2002) and nursery center (built in 2000) as 'village' and playground (constructed in 2002-) as 'square', made a proposal of 'growing farm'.

在院子里设计的水系边玩耍的孩子们

園庭につくられた水路であそぶこどもたち

Children play at a created stream in the kindergarten ground

爬攀工具

クライミング遊具

Climbing play structure

外观
園舎外観
External view of the school building

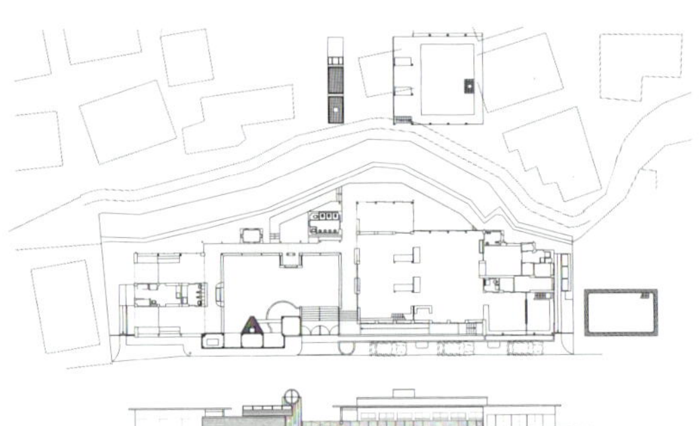

此项目摄影：大木宏之
撮影：大木宏之
Photo: Hiroyuki Oki

马尾保育院

所 在 地　　神奈川县横滨市
竣工时间　　2002 年 3 月
建筑结构　　钢结构 / 地上 1 层，地下 1 层
用地面积　　1192.33m²
占地面积　　506.82m²
建筑面积　　538.09m²

该保育院共招收园儿 60 名。在平面布局上将 0 － 1 岁儿和 2 － 5 岁儿的活动领域分为两大区域，并通过连廊将它们连接起来，而中庭则夹在这两个区域之间。在面向道路的中庭内设置了塔状的游戏玩具，并在 3 － 5 岁儿的活动室内布置了空中回廊和网状游戏装置。

塔状游具“马尾塔”
塔状遊具「つくしタワー」
'Tsukushi Tower'

园内
園舎内
Inside the school building

つくし保育園

所 在 地	神奈川県横浜市
竣 工 年	2002 年 3 月
規　　模	S 造／地上 1 階、地下 1 階
敷地面積	1,192.33m^2
建築面積	506.82m^2
延床面積	538.09m^2

園児数は 60 名、中庭をはさみ、0、1 才のブロックと2、3、4、5、才のブロックの 2 つに分節化し、それらは廊でつながれている。道路に面する中庭には塔状遊具が設けられ、3、4、5、才児室には空中回廊とネット遊具が設けられている。

Tsukushi Day Nursery

Location	Yokohama City, Kanagawa Prefecture
Completion	March 2002
Structure	Steel/1 floor below-grade, 1 above-ground floor
Site area	1,192.33m^2
Building area	506.82m^2
Total floor area	538.09m^2

Caring for 60 children, this day nursery has a central courtyard divided into two, with one section for ages over and under 1 year old and one section for ages for 2, 3, 4, and 5 years, and these are connected together by a corridor. There is tower play structure in the courtyard facing the road, and air corridor and net play equipment are furnished in the room for the children aged 3, 4 and 5 years.

若草保育院项目

所 在 地	东京都昭岛市
竣工时间	2003年9月
建筑结构	钢筋混凝土结构，部分为钢结构（屋顶部分）/地上2层，地下1层
用地面积	1436.61m²
占地面积	547.20m²
建筑面积	1125.43m²

该保育院位于东京都郊外的昭岛市的住宅区内，约有幼儿160人。在整体空间构成上，包括入口门厅部分和两座园舍。在设计上力图营造出有利于孩子们玩耍的空间氛围。

わかくさ保育園プロジェクト

所 在 地	東京都昭島市
竣 工 年	2003年9月
規　　模	RC造一部S造（屋根部）／地上2階、地下1階
敷地面積	1,436.61m²
建築面積	547.20m²
延床面積	1,125.43m²

東京都の郊外、昭島市の住宅街に立地する園児数約160人の保育園。エントランスホール部分と2つの園舎ブロックに分けられている。こどものあそびやすい空間構造理論にもとづいて全体が計画設計されている。

Wakakusa Day Nursery Project

Location	Akishima City, Tokyo
Completion	September 2003
Structure	RC, part Steel (Roof)/1 floor below-grade, 2 above-ground floors
Site area	1,436.61m²
Building area	547.20m²
Total floor area	1,125.43m²

A day nursery caring for about 160 children located in a residential area in Akishima City, in the suburbs of Tokyo. Divided into an entrance hall portion and two nursery school portions, it was planned and designed overall based on space structure theory in order to facilitate children's play.

多治见市泷吕小学校项目

所 在 地　岐阜县多治见市
竣工时间　2006 年 3 月
建筑结构　钢筋混凝土结构，部分为木结构，钢结构 / 地上 2 层，地下 1 层
用地面积　25816m²
占地面积　6470m²
建筑面积　10056m²

该学校的整体空间由位于南侧的校舍楼和北侧的体育馆以及屋顶上有游泳池的管理楼、在校舍楼和体育馆之间中庭构成。校舍楼被分为三个大的区域，并且每一个区域都各自包括有多功能空间和普通教室，从而形成了非常丰富的学习空间。出入口的设置也不同于以往的集中形式，而是为每一个教室都设置了出入口。

多治見市立滝呂小学校プロジェクト

所 在 地　岐阜県多治見市
竣 工 年　2006 年 3 月
規　　模　RC 造一部木造一部 S 造／地上 2 階、地下 1 階
敷地面積　25,816m²
建築面積　6,470m²
延床面積　10,056m²

南側に校舎棟、北側に中庭をはさんで体育館棟、屋上にプールがのる管理棟という構成。校舎棟は大きく3つのブロックに分かれ、それぞれが多目的スペースや普通教室をもち、豊かな学習空間を形成している。集中型の昇降口ではなく、各教室に昇降口が設けられているのが、極めて新しい提案である。

Takiro Elementary School Project

Location　Tajimi City, Gifu Prefecture
Completion　March 2006
Structure　RC, part W, Steel/1 floor below-grade, 2 above-ground floors
Site area　25,816m²
Building area　6,470m²
Total floor area　10,056m²

This project is composed of a classroom building section on the south side, a gym building section on the north side that straddles the courtyard, and an administrative building with rooftop pool. The classroom building block is divided into three primary blocks, and each possesses a multi-purpose space and ordinary classrooms, forming abundant space for study. Built with no centralized entrance way, but rather with entrances for each classroom, which is a totally new idea.

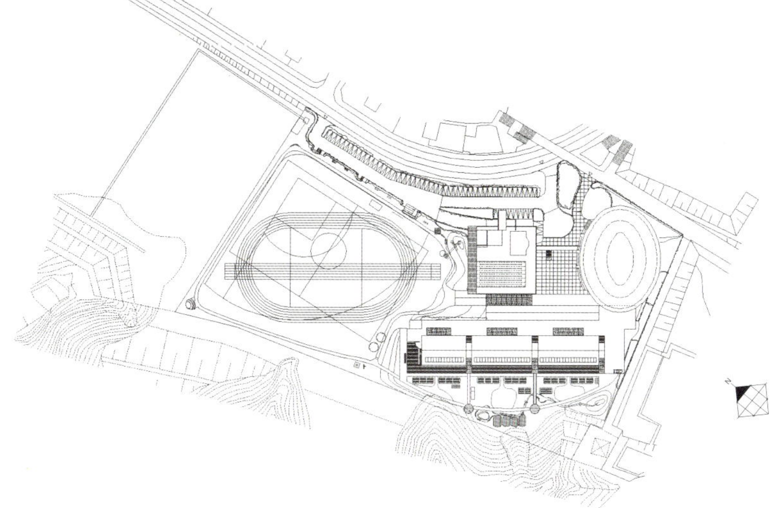

Housing

Welfare

住宅 / 医疗 · 福利设施

住宅 / 医療 · 福祉施設

HOUSING/WELFARE

从游泳池一边遥望相模湾

プールサイドより相模湾を望む

View of Sagami Bay from the swimming pool

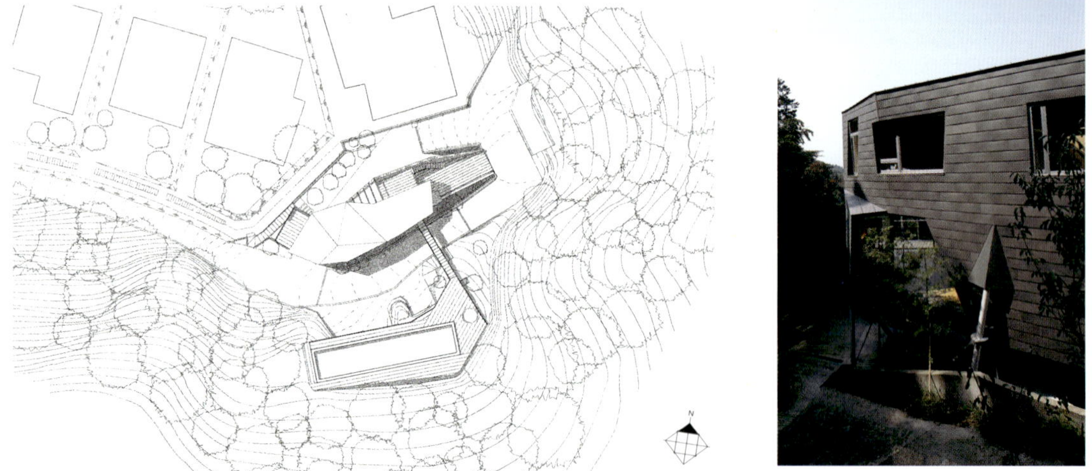

樱山之家

所 在 地	神奈川县逗子市
竣工时间	2001年7月
建筑结构	钢结构 / 地上3层
用地面积	3462 m²
占地面积	106.72 m²
建筑面积	194.5 m²
获奖经历	2003年日本建筑学会作品选集世入选

建筑用地位于逗子海岸附近的细长山谷深处,其周围是险峻的斜山坡,从用地的西北方向可以遥望到江之岛、相模湾和富士山。通往这里的道路却是非常的狭窄,并且需要攀登陡峭的山路。同时空气的高湿度也是必须考虑的因素,因此建筑的主要部分采用架空形式。在设计中充分利用106m²这个被限定了的建筑面积及其周边的自然环境,最大限度地实现建筑空间的丰富性,力图营造出能够使在这里生活的人们感觉到舒适和温馨的空间形态。

桜山の家

所 在 地	神奈川県逗子市
竣 工 年	2001年7月
規　　模	S造/地上3階
敷地面積	3,462 m²
建築面積	106.72 m²
延床面積	194.5 m²
受 賞 歴	2003.日本建築学会作品選集

敷地は逗子の海岸から近く、急峻な斜面に囲まれた細い谷戸の最奥部に位置し、北西の方向には江ノ島、相模湾、富士山を望むことができる。しかしアプローチが細く、極めて急勾配の道を登らねばならない。また谷戸のため高い湿気を考慮する必要があると考え、主要部分をピロティーにより持ち上げている。106平米という限られた建築面積と周辺の環境を利用して、空間の豊かさを最大限に獲得することと、ここで生活する人が気持ちの良い生活を送ることを可能とする空間形態を求めた。

Sakurayama House

Location	Zushi City, Kanagawa prefecture
Completion	July 2001
Structure	Steel/3 above-ground floors
Site area	3,462 m²
Building area	106.72 m²
Total floor area	194.5 m²
Awards	2003 Architectural Institute of Japan Selected Architectural

The site is located on the Zushi coast and located in the deepest part of a narrow moor, surrounded by steep hillsides, with Enoshima, Sagami Bay and Mt. Fuji viewable to the northwest. However, the approach is narrow and it is necessary to climb up an extremely steep road. Additionally, since it is moorland, it was necessary to consider the high humidity, so a major portion was supported by pilotis. Using the limited construction area of 106sqm and the surrounding environment, a spatial form was sought in which an abundance of space could be acquired as fully as possible and a comfortable life could be had for the persons who live here.

北侧外观，建立在有落差的不规则形状的用地上
北側からの外観.不整形で段差のある敷地に建つ
Northern elevation. The building is located at an irregular shaped site with different levels

南侧外观的夜景

南側外観の夜景

Southern facade at night

连接一层和二层的楼梯
1階と2階を結ぶ階段
Staircases connecting the first and second floors

从和室内可以穿越食堂一直看到起居室，室内地板使用的是线形无垢材料，顶棚使用的是无框无垢材料

和室から食堂越しに居間までを見通す.床はクリ無垢材、天井はキリ無垢材を使用

View of living room from Japanese style room through kitchen. Chestnut timbers are used for the floor and paulownia timbers are used for the ceiling

从三层看二层的起居室

３階から２階居間を見る

Viewing living room on the 2nd floor from the 3rd floor

轻井泽 640

所 在 地	长野县北佐久郡
竣工时间	1984 年 6 月
建筑结构	木结构 / 地上 2 层
用地面积	2457.32 m²
占地面积	421.79 m²
建筑面积	540.47 m²

该建筑是某企业为员工提供的疗养设施，设计条件是7个家庭或集体在1年当中随时都可以利用，并需要设置一个网球场。在轻井泽的自然环境当中以怎样的形式设计具有集合住居性格的住宅建筑成为本设计的焦点。最后决定将主要面向家庭的跃层式的3组住宅和1、2层主要面向团体的单元式的2组住宅并列布置在用地深处的地势较高的位置上。并且将全体建筑作为一个家来进行设计。

軽井沢 640

所 在 地	長野県北佐久郡
竣 工 年	1984 年 6 月
規　　模	木造／地上 2 階
敷地面積	2,457.32 m²
建築面積	421.79 m²
延床面積	540.47 m²

この施設はある企業の社員のための保養施設で、計画の条件は7つの家族もしくはグループが一年を通じ利用できることと、テニスコートを設けることであった。軽井沢の自然の中で、集合住居的性格の建物のありかたを考慮することが課題となった。その結果、主に家族向けのメゾネットタイプの3ブロックと、1、2階に設けたグループ向けのフラットタイプの2つのブロックを敷地の奥の高いところに一列に配置した。さらに全体にポーラス（多孔質）な形態を持つ一つの大きな家とした。

Karuizawa 640

Location	Kitasaku-gun, Nagano prefecture
Completion	June 1984
Structure	W/2 above-ground floors
Site area	2,457.32 m²
Building area	421.79 m²
Total floor area	540.47 m²

This facility is intended as a resort for company employee, requiring provision for simultaneous use by seven families or groups, and tennis courts, and aiming to fit into Karuizawa's natural environment. The result was three blocks of mainsonettes for families and two blocks of mainsonettes for families and two blocks of one- and two-story flats for groups in the form of one large porous building located at the back of the site.

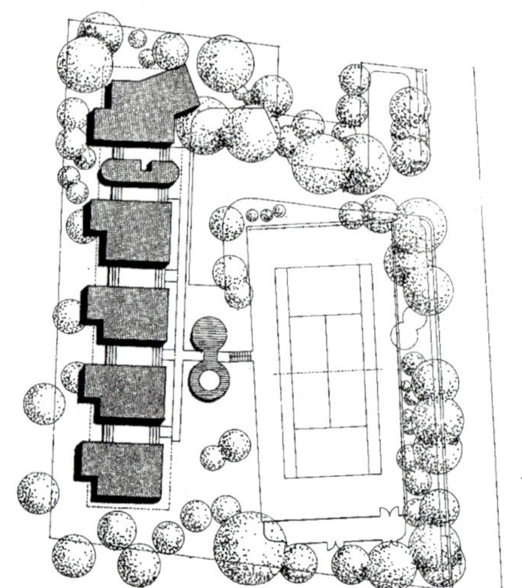

南侧外观，外墙壁贴的是落叶松板材，前面是网球场

南側外観.外壁はカラマツ板張り.手前にはテニスコートが広がる

Southern elevation. Larch timbers are used for the exterior wall. There is a tennis court on the front

跃层式A楼内装修均有所不同
メゾネットタイプA棟. タイプ毎にインテリアが異なる
Maisionette type A. Different types have different interiors

跃层式 C 栋
メゾネットタイプ C 棟
Maisionette type C

同层式 D 栋
フラットタイプ D 棟
Flat type D

跃层式 C 栋
メゾネットタイプ C 棟
Maisionette type C

秋田县营住宅御野庭团地

所 在 地	秋田县秋田市
竣工时间	1981 年 4 月
建筑结构	钢筋混凝土结构 / 地上 4 层
用地面积	13628 m^2
占地面积	3078 m^2
建筑面积	8922 m^2
获奖经历	1982 年秋田市城市景观奖

因为建筑用地狭窄细长，长边方向为南北向，所以住户的日照问题可以得到保证。在平面布局中设计了一个中庭，并将中庭的地面设计出和缓的起伏，同时为孩子们提供了一个游戏的空间。在设计手法上充分借鉴了秋田当地传统民居的地方特色。

秋田県営住宅御野庭団地

所 在 地	秋田県秋田市
竣 工 年	1981 年 4 月
規　　模	RC 造／地上 4 階
敷地面積	13,628 m^2
建築面積	3,078 m^2
延床面積	8,922 m^2
受 賞 歴	1982.秋田市都市景観賞

秋田の伝統的な白壁の切妻町屋の妻入り連続ファサードを引用。敷地は細長く、長辺方向が南北を向いている。そのため住戸の日照の確保を考え、中庭を設けた。そこには緩やかな起伏とこども達のあそび場を設けた。

Akita Prefectural Onoba Housing

Location	Akita City, Akita prefecture
Completion	April 1981
Structure	RC/4 above-ground floors
Site area	13,628 m^2
Building area	3,078 m^2
Total floor area	8,922 m^2
Awards	1982 Akita Urban Landscape Award

Gables including serial facades of the traditional Akita white wall gabled roof town house were employed. The site was narrow and long, with long side facing north and south, thus, to secure illumination for living, an interior courtyard was created, and inside this space a moderately rolling terrain and a children's play space were fashioned.

从西北角看屋顶的连接
北西から見た屋根のつながり
Viewing roofs from north west

从楼梯处看中庭

住棟階段から中庭を見る

Viewing courtyard from staircase

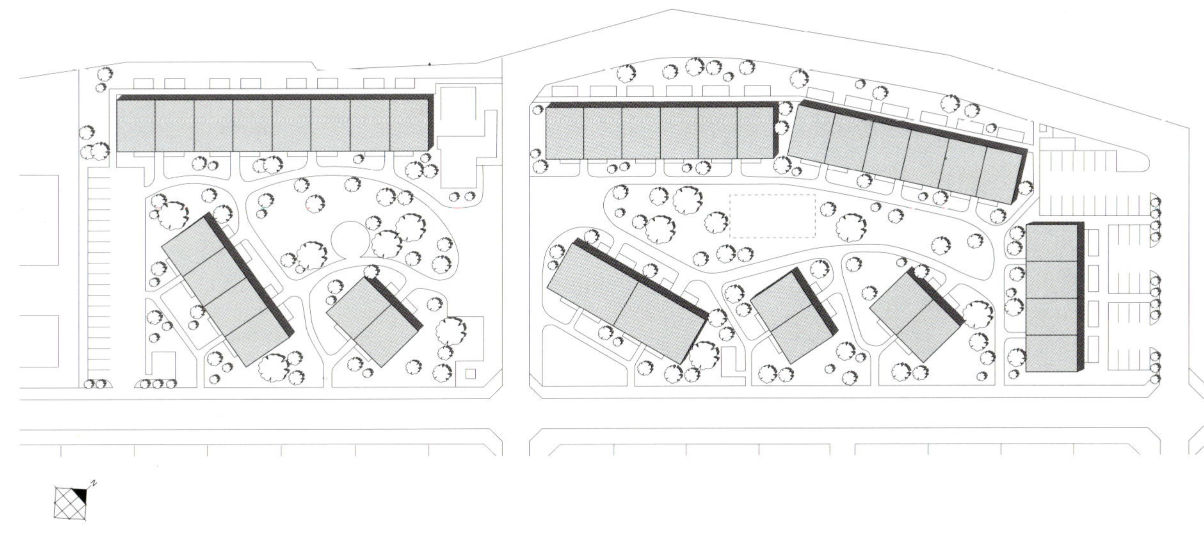

屋顶泳池
屋上プール
Roof top swimming pool

片濑山的家

所 在 地	神奈川县藤泽市
竣工时间	1980年3月
建筑结构	钢筋混凝土结构/地上3层
用地面积	599 m²
占地面积	77 m²
建筑面积	152 m²

该住宅为建筑师自己的府邸。建筑用地约600m²，其中1/3是平地，其余的都是坡地。建筑物就建在用地的平坦部分。结构形式采用钢筋混凝土结构。屋顶上是10m × 3m的游泳池和约25m²的屋顶平台。从书房通过木制楼梯可以上到屋面。

片瀬山の家

所 在 地	神奈川県藤沢市
竣 工 年	1980年3月
規　　模	RC造／地上3階
敷地面積	599 m²
建築面積	77 m²
延床面積	152 m²

建築家の家である（自邸）。敷地は約600㎡で、そのうち平地が1/3、残りは斜面である。建物は平地の部分2階建コンクリート造でつくられた。屋上は10m × 3mのプールと約25㎡のテラスになっており、書斎より木製の階段でつながっている。

Kataseyama House

Location	Fujisawa City, Kanagawa prefecture
Completion	March 1980
Structure	RC/3 above-ground floors
Site area	599 m²
Building area	77 m²
Total floor area	152 m²

This is the architect's own residence. The site is 600m², of which one-third are flat, and the rest sloping. The structure is two-story concrete construction on the level part of the site. On the roof is a small (3x10m) pool and 25m² terrace, connected to the study by stairs of wood.

2层起居室
2階居間
Living room on the second floor

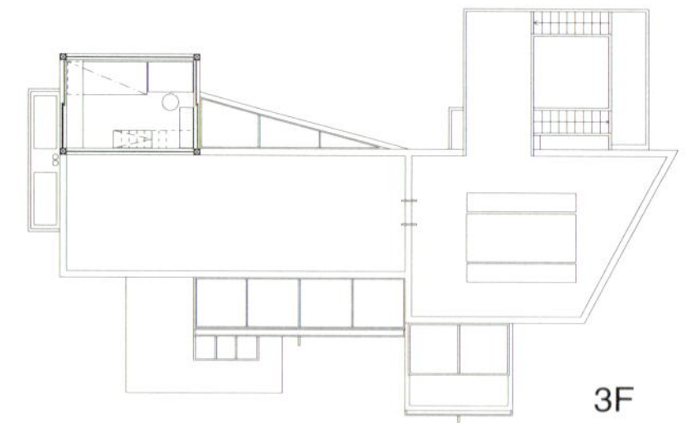
3F

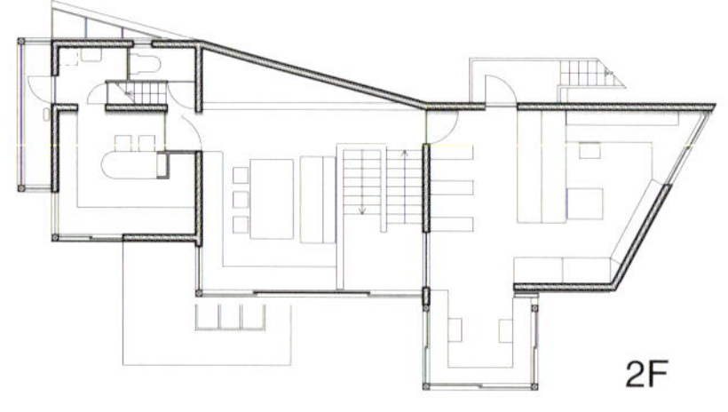
2F

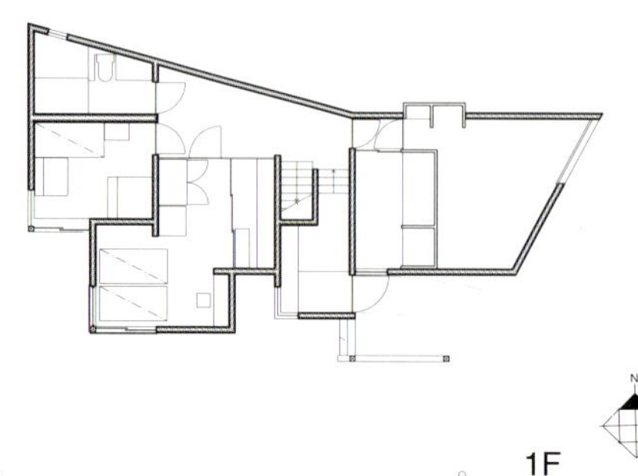

1F

全景
全景
External view of the house

轻井泽C山庄

所 在 地	长野县北佐久郡
竣工时间	1989年4月
建筑结构	本结构，部分为钢框架钢筋混凝土结构/地上2层，地下1层
用地面积	988.88 m²
占地面积	197.36 m²
建筑面积	240.55 m²

本山庄建于日本最具代表性的别墅区轻井泽。因为建筑的平面形状与C型相似，所以人们在室内踱步就可以欣赏到建筑空间和周围自然景观的变化。

軽井沢C山荘

所 在 地	長野県北佐久郡
竣 工 年	1989年4月
規　　模	木造一部SRC造／地上2階、地下1階
敷地面積	988.88 m²
建築面積	197.36 m²
延床面積	240.55 m²

日本の代表的な別荘地である軽井沢に建つ山荘である。建物の平面がC型に似た形をしているため、人々は室内を歩き回ることによって空間と景観の変化を楽しむことができる。

Mountain Villa 'C', Karuizawa

Location	Kitasaku-gun, Nagano prefecture
Completion	April 1989
Structure	W, part SRC/1floor below-grade, 2 above-ground floors
Site area	988.88 m²
Building area	197.36 m²
Total floor area	240.55 m²

This villa stands in one of the most representative mountain resorts in Japan, Karuizawa. Being more or less like a letter "C" on plan, this makes it possible to enjoy the trees from various parts of the interior and also makes for rather interesting collection of interior spaces.

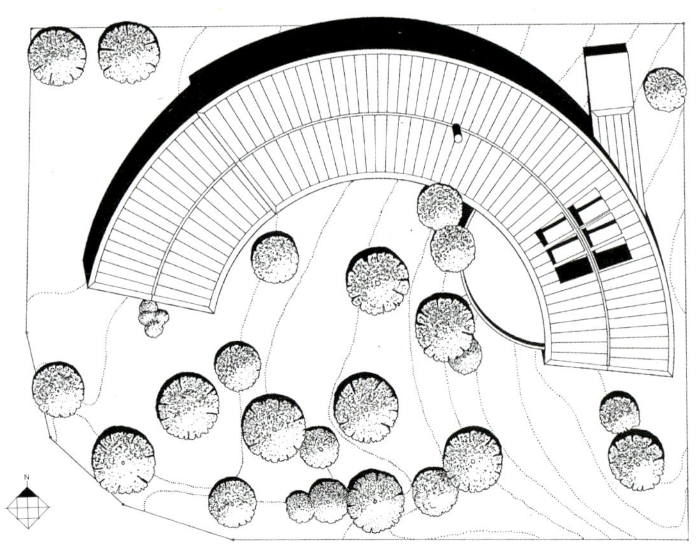

2层主屋

母屋 2 階

2nd floor of the main house block

玻璃屋顶的大厅，右侧是起居室，左侧是和室，正面是南侧的露台

ガラス屋根のロビー．右側は居間、左側は和室、正面は南側テラス

Glass-roofed lobby with living room to the right, Japanese style room to the left and terrace to the south

起居室，可以看到屋顶构架

居間を望む.小屋組がそのまま見える

View of the living room, showing the exposed rafters

鹄沼的家

所 在 地	神奈川县藤泽市
竣工时间	1990 年 4 月
建筑结构	木结构 / 地上 2 层
用地面积	147.28 m²
占地面积	73.35 m²
建筑面积	117.57 m²
获奖经历	1990 年度神奈川县建筑奖

本建筑是建于海滨城市住宅区中的小住宅。建筑平面呈变了形的十字型，在设计中尽可能地在建筑的周边布置大片的绿地。起居室和厨房布置在2层，个人房间、卧室和浴室布置在1层。

鵠沼の家

所 在 地	神奈川県藤沢市
竣 工 年	1990 年 4 月
規　　模	木造／地上 2 階
敷地面積	147.28 m²
建築面積	73.35 m²
延床面積	117.57 m²
受 賞 歴	1990.神奈川建築コンクール

海浜都市の住宅地に建つ小住宅である。変形した十字型平面を持ち、周辺にできるだけ多くの緑を配した。2階は居間・台所のワンルーム形式で、1階に個室、寝室、浴室がある。

Kugenuma House

Location	Fujisawa City, Kanagawa Prefecture
Completion	April 1990
Structure	W/2 above-ground floors
Site area	147.28 m²
Building area	73.35 m²
Total floor area	117.57 m²
Awards	1990 Kanagawa Architecture Award

This small house is located in a residential area of a seaside city. Seen as a possible answer to building a small house in cramped urban conditions, the building is cruciform in plan, and there is as much planting as could be sensibly fitted into the rest of the site. Additionally, the combined living room-cum-kitchen is upstairs, while the more private bedrooms and bathroom are at ground level.

起居室，食堂，厨房是一个约15坪的整体空间，顶棚呈HP贝壳形状，从南侧坡向北侧

居間、食堂、厨房は約15坪の一体的空間. 天井はHPシェルになっており、南側から北側に低くなる

The living room, dining room and kitchen form an integrated space of around 50sqm. The ceiling, which slopes from south to north, is an HP shell

松庵的家

所 在 地	东京都杉并区
竣工时间	1990 年 12 月
建筑结构	钢筋混凝土结构 / 地上 3 层
用地面积	232.48 m²
占地面积	108.63 m²
建筑面积	232.18 m²

本建筑是位于东京都杉并区松庵内的住宅建筑，平面形状呈十字型。因为在建筑的四个角落都布置了绿地，所以不论在室内的哪个房间都可以欣赏到绿色，而且也能够阻挡周围的视线。

松庵の家

所 在 地	東京都杉並区
竣 工 年	1990 年 12 月
規　模	RC 造／地上 3 階
敷地面積	232.48 m²
建築面積	108.63 m²
延床面積	232.18 m²

東京の住宅地である杉並区の松庵につくられた 70 坪の住宅である。平面は十字型をし、四角よすみに緑を配している。こうすることによって、どの室からも緑を楽しむことができ、また、隣地との視覚的距離も確保できる。

Shoan House

Location	Suginami-ku, Tokyo
Completion	December 1990
Structure	RC/3 above-ground floors
Site area	232.48 m²
Building area	108.63 m²
Total floor area	232.18 m²

This house stands in the predominantly residential district of Suginami Ward in Tokyo. It is cruciform on plan with planting in the four open corners of the site, meaning that views of the garden can be enjoyed from every room, while visual distance from the adjoining sites can be secured.

两侧外观全貌，为了确保环境的舒适，在道路的两旁也布置了小型的公园绿地

西側外観全景.アメニティ確保のための道路側にもポケットパークが設けられている.

View of the west face and pocket size garden along the amenity road

工作室

アトリエ

Atelier

食堂

食堂

Dining room

群星艺术 23

所 在 地	神奈川县横滨市
竣工时间	1992 年 4 月
建筑结构	钢结构 / 地上 2 层
用地面积	893.39 m^2
占地面积	283.89 m^2
建筑面积	412.73 m^2
获奖经历	1993 年度商业环境设计奖

因为业主是艺术家，所以他的家既是展室，工作室，又是个人的私人住宅。带有大吹拔空间的展室和创作室以及起居室等空间通过坡道和楼梯连接起来，从而形成了一个环游的空间结构形态。

スターズ・アート 23

所 在 地	神奈川県横浜市
竣 工 年	1992 年 4 月
規 模	S 造／地上 2 階
敷地面積	893.39 m^2
建築面積	283.89 m^2
延床面積	412.73 m^2
受 賞 歴	1993.商環境デザイン賞

施主がアーティストであるこの家は、作家のギャラリーであり、スタジオであり、個人的なすまいでもある。大きな吹き抜けのあるギャラリー、創作室、ジャグジーのある居室等がスロープとスリリングな階段によってつなげられ、空間構成として遊環構造が形成されている。

Stars Art 23

Location	Yokohama City, Kanagawa prefecture
Completion	April 1992
Structure	Steel/2 above-ground floors
Site area	893.39 m^2
Building area	283.89 m^2
Total floor area	412.73 m^2
Awards	1993 JCD Award

This house, built for an artist, comprises gallery space, a studio as well as private accommodation. The gallery with its high ceiling, workroom and living areas where there is a Jacuzzi are linked by a slope and excitingly planned stairs, making the whole spatial composition a true piece of circular play system architecture.

从连廊就可以看到突塔的下部，象征性地设置了用玻璃围合而成的圆形房间
ギャラリーから見える突塔下部.ガラスで囲まれた円形の部屋が象徴的に設けられている
Circular glazed room beneath the spire seen from the gallery

东侧全貌
東側全景
General view of the east

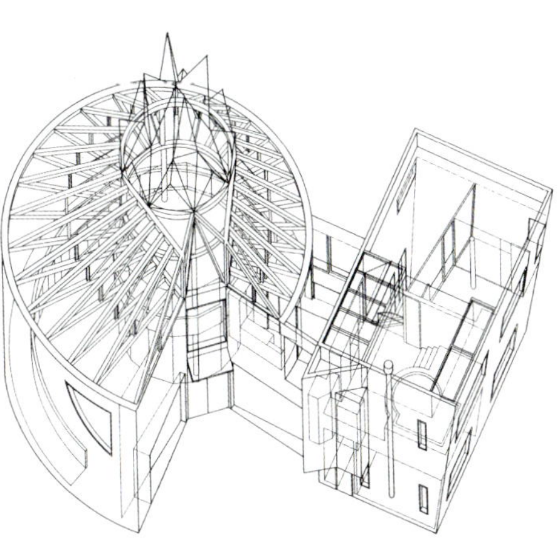

春日部的家

所 在 地	琦玉县春日部市
竣工时间	1998年3月
建筑结构	主房木结构（地上1层）；客房钢结构，部分为木结构（地上2层）
用地面积	1576.62 m^2
占地面积	480.95 m^2
建筑面积	603.63 m^2
获奖经历	1999年优良设计奖

本建筑物是为三代人、两个家庭所设计的住宅。该住宅在整体空间形态设计上既体现了由主房、客房、门房围合中庭布置的传统的空间构成形式，同时又试图营造出一种具有现代气息的舒适的居住生活空间环境。

春日部の家

所 在 地	埼玉県春日部市
竣工年月	1998年3月
規　　模	母屋木造（地上1階）・客屋 S造・木造（地上2階）
敷地面積	1,576.62 m^2
建築面積	480.95 m^2
延床面積	603.63 m^2
受 賞 歴	1999.グッドデザイン賞

この建物は3世代、2世帯のための住宅である。庭を囲む母屋、客屋、門屋の構成に伝統的なたたづまいをもちながら、全体としては現代的で快適で楽しさのある住生活が営まれるような空間構成を意図している。

Kasukabe House

Location	Kasukabe City, Saitama Prefecture
Completion	March 1998
Structure	W (1st floor), Steel and W (2nd floor)
Site area	1,576.62 m^2
Building area	480.95 m^2
Total floor area	603.63 m^2
Awards	1999 Good Design Award

A building for three generations and two families, composed of a main house, guest house and gate house surrounding a garden with a traditional atmosphere, but as a whole a building in which modern, comfortable and enjoyable living can be operated was intended.

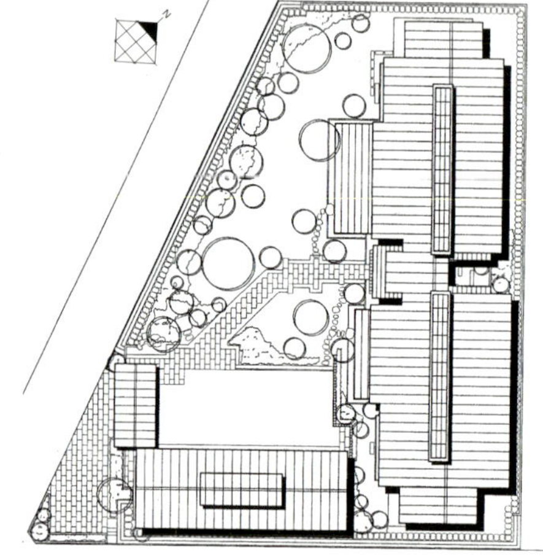

主房室内的接待室

母屋内部応接室

Living room inside the main house

主房南立面
門屋より母屋南面を望む
Southern elevation of the main house seen from the gate house

从日式房间看起居室
和室から居間を見る
Viewing living room from Japanese style room

主屋玄关
母屋玄関
Entrance hall of the main house block

香蕉之家

所 在 地	山梨县南都留郡
竣工时间	1982 年 1 月
建筑结构	木结构 / 地上 2 层
用地面积	796.8 m²
占地面积	117 m²
建筑面积	93 m²

建筑用地为 30 度倾斜坡地。住宅平面沿着用地的等高线方向进行布置，因此形成香蕉形状。从起居室的正前方可以遥望到富士山。

バナナハウス

所 在 地	山梨県南都留郡
竣 工 年	1982 年 1 月
規　　模	木造／地上 2 階
敷地面積	796.8 m²
建築面積	117 m²
延床面積	93 m²

敷地は南斜面で約30度の斜度を持つ。平面形は敷地の等高線をそのままなぞり、バナナの形になった。居間からは真正面に富士山を望むことができる。

Banana House

Location	Minami-Tsuru-gun, Yamanashi prefecture
Completion	January 1982
Structure	W/2 above-ground floors
Site area	796.8 m²
Building area	117 m²
Total floor area	93 m²

The lot faces south on a 30° slope. The outline of the house resembles a banana, hence the name of the house. Mt. Fuji can be seen directly from the living room.

从入口处俯瞰，香蕉形的曲面墙

アプローチより下を見る.バナナの形のように湾曲する壁面

Viewing down from the approach. The wall curves like shape of banana

玄关

玄関

Entrance hall

从二层的平台俯瞰起居室

２階デッキから居間を望む

Viewing living room from deck on the 2nd floor

从二层的起居室可以看到富士山

２階居間の窓からは富士山が見える

It is able to view Mt. Fuji from window of the second floor living room

大森的家

所 在 地	东京都大田区
竣工时间	1999 年 2 月
建筑结构	钢筋混凝土结构 / 地上 3 层
用地面积	122.66 m^2
占地面积	70.77 m^2
建筑面积	192.48 m^2

因为主人准备与父母一同生活，因此将该住宅设计成两代人共同居住型的住宅。主人是一位 30 岁出头的年轻人，两个孩子也尚年幼。住宅的 1 层是父母的房间，2 层是起居室、餐厅和厨房等公共部分。3 层是卧室层，设有主卧室和儿童房。在屋顶设计了一个 2.5m × 10m 的游泳池。

大森の家

所 在 地	東京都大田区
竣工年月	1999 年 2 月
規　　模	RC 造 地上 3 階
敷地面積	122.66 m^2
建築面積	70.77 m^2
延床面積	192.48 m^2

建主は30代前半と若く、ふたりのこどもも幼児である。ご主人の両親と住む予定で2世帯住宅として計画された。1階に両親の部屋、2階はリビングとダイニング、キッチンというこの住宅のパブリックな部分である。3階は寝室階で、主寝室とこども部屋がある。屋上には 2.5 m× 10 mのプールを設けた。

Omori House

Location	Ota-ku, Tokyo
Completion	February 1999
Structure	RC/3 above-ground floors
Site area	122.66 m^2
Building area	70.77 m^2
Total floor area	192.48 m^2

Clients were young, in early 30s with two infant children, and planned to live with the paternal in-laws, so this house was planned as two-family residence. The first floor contains rooms for the in-laws and the second floor is public space for the residence--the living room, dining room and kitchen. The third floor is the bedroom floor, with the master bedroom and children's bedroom. A 2.5m by 10m pool is located on the roof.

屋顶的游泳池
屋上に設けられたプール
Rooftop swimming pool

东侧全景

東側からの全景

General view of the east

由餐厅看起居室

ダイニングよりリビング方向を見る

Living room seen from dining room

由卧室穿过中庭看儿童房

寝室より吹き抜けを介して子ども部屋を見る

Children's room seen from bedroom through atrium

末吉河畔家园

所 在 地	冲绳县那霸市
竣工时间	1988 年 3 月
建筑结构	钢筋混凝土结构
用地面积	东楼 559.09 m² 西楼1311.29 m²
占地面积	东楼223.58 m² 西楼524.05 m²
建筑面积	东楼 437.56 m² 西楼1027.14 m²
获奖经历	1989 年那霸市城市景观奖

该住宅建筑是为30-40岁年龄层的家庭设计的销售住宅。分两栋，共有16套住居，分别是由5户和11户所组成的连排式住宅。

末吉リバーサイドテラス

所 在 地	沖縄県那覇市
竣 工 年	1988 年 3 月
規　　模	RC 造
敷地面積	東棟 559.09 m² 西棟 1,311.29 m²
建築面積	東棟 223.58 m² 西棟 524.05 m²
延床面積	東棟 437.56 m² 西棟 1,027.14 m²
受 賞 歴	1989.那覇市都市景観賞

末吉リバーサイドテラスは30～40代の家族のための分譲住宅として計画された。全体として16戸であるが、2棟にわかれており、5戸と11戸の連棟式タウンハウスである。

Sueyoshi Riverside Terrace

Location	Naha,Okianwa Prefecture
Completion	March 1988
Structure	RC
Site area	559.09 m² (East Block); 1,311.29 m² (West Block)
Building area	223.58 m² (East Block); 524.05 m² (West Block)
Total floor area	437.56 m² (East Block); 1,027.14 m² (West Block)
Awards	1989 Naha Urban Landscape Award

Sueyoshi Riverside Terrace was designed as townhouse living for families of couples in their thirties and forties. There are altogether 16 homes in two wings, five in one and eleven in the other.

铺设红瓦的屋顶
赤瓦葺きしっくい固めの屋根
Red tiled roof using plaster

从屋顶小屋俯瞰起居室
屋根裏部屋から居間を望む
Looking down into the living room from the attic space

坐落在落叶松林中，可以看到建筑顶部的1.5m见方的展望台兼书房
カラマツの林の中に建つ 建物トップに1.5m角の展望台兼書斎がみえる
Located in a larch forest, the 1.5 meter square study or "rookery" top the building

鸟之家

所 在 地	长野县南佐久郡
竣工时间	1990年11月
建筑结构	木结构 / 地上1层
占地面积	2427 m²
建筑面积	71 m²

该建筑位于八岳山的半山腰，从室内就可以看到八岳山的山顶。因其平面形状与张开翅膀的小鸟相似，因此取名〝鸟之家〞。

鳥の家

所 在 地	長野県南佐久郡
竣 工 年	1990年11月
規　　模	木造／地上1階
敷地面積	2,427 m²
延床面積	71 m²

八ヶ岳の中腹にたてられた山荘である。3階部分は1.5m四方のガラス張りの書斎である。この部屋から八ヶ岳の山頂を望むことができる。平面形が鳥の羽を広げたような形から、"鳥の家"と名づけられた。

Bird House

Location	Minami Saku-gun, Nagano Prefecture
Completion	November 1990
Structure	W/1 above-ground floor
Site area	2,427 m²
Total floor area	71 m²

This mountain cabin built on the slopes of the range of eight peaks, "Yatsugatake". The "rookery" is a 1.5 meter square study at third floor level, with glass on all sides, from where there are wonderful views of Yatsugatake. The house was named "Bird House" because of the building shape as if a bird is spreading the wings.

白金台园亭

所 在 地	东京都港区
竣工时间	1989 年 10 月
建筑结构	钢筋混凝土结构/4层地上，地下1层
用地面积	699.49 m^2
建筑面积	1221.25 m^2

白金台地处东京城市中心的高级住宅区，它与目黑区的自然学习园相邻，周围环境十分幽静。本次设计的内容是将原有的私人住宅改造成为地上3层、地下1层的公共住宅。

白金台ガーデンハウス

所 在 地	東京都港区
竣 工 年	1989 年 10 月
規 模	RC 造／4 階地上、地下 1 階
敷地面積	699.49 m^2
延床面積	1,221.25 m^2

都心部の高級住宅地である白金台で、隣接して目黒の自然学習園がある静かな環境に位置する。個人の家を建て替え、地下1階地上3階の共同住宅とした。

Shiroganedai Garden House

Location	Minato-ku, Tokyo
Completion	October 1989
Structure	RC/1floor below-grade,4 above-ground floors
Site area	699.49 m^2
Total floor area	1,221.25 m^2

Here in Shiroganedai, one of the better residential areas within the metropolis private dwelling, was demolished to make way for a housing development comprising two blocks, each with one basement and three storeys above ground.

出租楼外观，穿过门就是夹在两栋楼之间的中庭，从这里就可以进入到各个住户

賃貸棟外観．門をくぐると2つの棟に挟まれた中庭があり、ここから各住戸にアプローチする

External view of a block rented housing. The inner courtyard, which is situated between two buildings leads to each individual dwelling

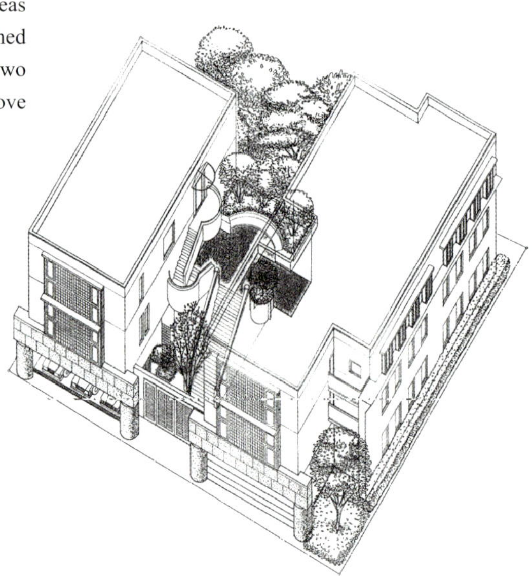

入口
エントランス
Entrance

元麻布集合住宅

所 在 地	东京都港区
竣工时间	1988 年 5 月
建筑结构	钢筋混凝土结构，部分为钢结构/地上 4 层，地下 1 层
用地面积	560.42 m^2
建筑面积	1409.24 m^2

该集合住宅位于使馆区的高级住宅小区内，设计的主要内容是将原有的私人住宅改建成为租赁住宅。楼梯设置在中央部分，不仅便于交通，而且公共空间显得十分宽敞明亮。

元麻布集合住宅

所 在 地	東京都港区
竣 工 年	1988 年 5 月
規　模	RC 造一部 S 造／地上 4 階、地下 1 階
敷地面積	560.42 m^2
延床面積	1,409.24 m^2

この建物は大使館等がある高級住宅地にある。個人住宅を建て替え、賃貸用の集合住宅として計画された。中央に階段を配し、明るいパブリックな空間を演出した。

Motoazabu Housing

Location	Minato-ku, Tokyo
Completion	May 1988
Structure	RC, part Steel/1floor below-grade, 4 above-ground floors
Site area	560.42 m^2
Total floor area	1,409.24 m^2

Here too, a private dwelling was demolished to make way for a piece of rentable collective housing, this time in a high-class residential area where there are several embassies. A light public space was achieved by placing the stairs centrally.

起居室
居間
Living room

卧室
寝室
Bedroom

由比之滨的家

所 在 地	神奈川县镰仓市
竣工时间	1995年4月
建筑结构	木结构／地上2层
用地面积	124.54 m^2
占地面积	49.6 m^2
建筑面积	98.35 m^2

该住宅位于距离镰仓市的由比之滨200m远的住宅区内。1层是个人房间，2层是起居室、食堂和日式房间，而卫生间和浴室就布置在楼梯平台上。

由比ヶ浜の家

所 在 地	神奈川県鎌倉市
竣 工 年	1995年4月
規 模	木造／地上2階
敷地面積	124.54 m^2
建築面積	49.6 m^2
延床面積	98.35 m^2

鎌倉の由比ガ浜の海岸から200mほど入った住宅地にこの家は立地している。1階は個室、2階に居間・食堂・和室があり、それをつなぐ踊り場にトイレ・浴室が配されている。

Yuigahama House

Location	Kamakura City, Kanagawa Prefecture
Completion	April 1995
Structure	W/2 above-ground floors
Site area	124.54 m^2
Building area	49.6 m^2
Total floor area	98.35 m^2

This house is located in a residential area 200m from the seacoast of Yuigahama in Kamakura. On the first floor are the private rooms and on the second floor are the living room, dining room and Japanese style rooms, with the toilet and bathroom arranged on the connecting landing.

在入口处安排绿化，减少压迫感
エントランス廻りには植栽を配し、圧迫感をやわらげている
Planting around the entrance soften oppressive feelings

起居室、食堂
居間・食堂
Living and dining room

伊势崎市营住宅

所 在 地	群马县伊势崎市
竣工时间	2005 年 3 月
建筑结构	钢框架钢筋混凝土结构，部分为钢筋混凝土结构、钢结构 / 地上 13 层
用地面积	2990 m²
占地面积	1642.70 m²
建筑面积	5300.79 m²

该建筑为 13 层混合住宅，1 层是保育园，1 层以上是住宅，可容纳 50 户。1 层部分的保育园可招收约 60 名幼儿，在建筑用地的北侧是可停放 50 台车辆的两层停车场。作为站前地区再开发计划的重要一环，同时为了实现开发商希望将该住宅出售给年轻家庭的愿望，因此在设计上力图体现该建筑在站前地区的标志性和充满魅力的空间构成。

伊勢崎市営住宅プロジェクト

所 在 地	群馬県伊勢崎市
竣 工 年	2005 年 3 月
規 模	SRC造一部RC造、S造／地上13階
敷地面積	2,990 m²
建築面積	1,642.70 m²
延床面積	5,300.79 m²

1 階に保育園、上部に 50 戸の住宅を持つ 13 階建ての集合住宅プロジェクト。1 階部分の保育園は園児数約 60 名、北側に約 50 台の 2 層駐車場を持つ。駅前再開発の一環としてシンボル性があり、かつ若い世帯が入居を希望するような魅力ある空間構成をもつ集合住居を提案している。

Isezaki Public Housing Project

Location	Isezaki City, Gunma Prefecture
Completion	March 2005
Structure	SRC, part RC and Steel/13 above-ground floors
Site area	2,990 m²
Building area	1,642.70 m²
Total floor area	5,300.79 m²

A housing project of 13 stories, with a day nursery on the first floor and 50 residences above. The first floor day nursery can care for about 60 children, and on the north side there is parking on two levels for about 50 cars. The project has a symbolic character as a link in the redevelopment of the station area, and the approach is of housing with attractive spatial structure to attract young families to take up residence there.

国立成育医疗中心

所 在 地	东京都世田谷区
竣工时间	2001 年 11 月(病房楼)、2003 年 3 月(成育庭园)、2004 年（研究楼）
建筑结构	钢框架钢筋混凝土结构／地上 12 层，地下 2 层
用地面积	75469.77 m^2
占地面积	9661.94 m^2 (病房楼)
建筑面积	64662.00 m^2 (病房楼)
获奖经历	2002 年 SDA 奖，展示设计奖，展示产业奖
共同设计	厚生省国立病院部经营指导科，日建设计

本中心是国守人藏医院和国立儿童医院合并而成的，是从妊娠到出生、幼儿、青春期一直到成人的连贯制治疗的医疗实践机构。考虑到游戏在儿童患者的治疗以及恢复当中的重要作用，因此提出了以病房楼、外来以及成育庭园为中心的拥有游戏功能的环境设计的实施方案。

国立成育医療センター

所 在 地	東京都世田谷区
竣 工 年	2001 年 11 月（病棟）、2003 年 3 月(成育庭園)、2004 年（研究棟）
敷地面積	75,469.77m^2
建築面積	9,661.94m^2（病棟）
延床面積	64,662.00m^2（病棟）
共同設計	厚生省国立病院部経営指導課、日建設計
受 賞 歴	2002.SDA 賞、ディスプレイデザイン賞、ディスプレイ産業賞

国立大蔵病院と国立小児病院が統合され、妊娠から出生、小児、思春期、成人へという総合的、継続的な医療を実践する国立成育医療センターとして生まれた。病気のこども達の治療、回復には遊びが重要な役割を果たすとの考えにもとづき、病棟、外来及び成育庭園を中心にあそび機能をもった癒しの環境デザインを提案し設計した。

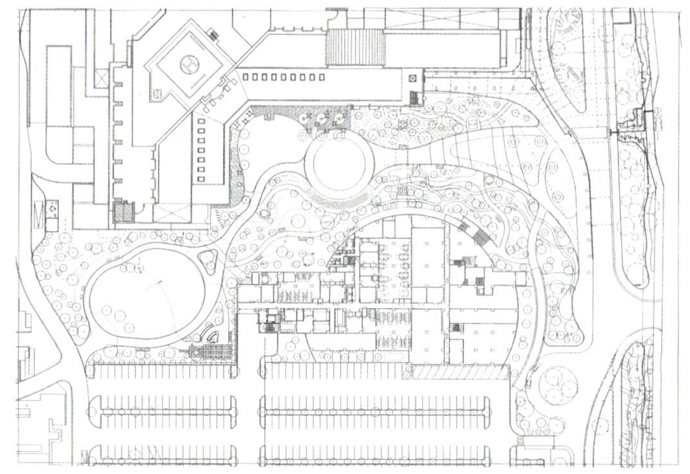

National Center for Child Health and Development

Location	Setagaya-ku, Tokyo
Completion	November 2001 (Hospital Block); March 2003 (Healing Garden); 2004 (Laboratory Block)
Structure	SRC/2 floor below-grade, 12 above-ground floors
Site area	75,469.77m^2 (Hospital Block)
Building area	9,661.94m^2 (Hospital Block)
Total floor area	64,662.00m^2 (Hospital Block)
Co-operative Design	Ministry of Health and Welfare Dept. of National Hospitals, Nikken Sekkei
Awards	2002 SDA Award, Display Design Award, Display Industry Award

This National Center for Child Health and Development was planned by unification of National Okura Hospital and National Children's Hospital for realizing holistic and continuous medical treatment from pregnant, birth, children, adolescent, to adult stage. Based on concepts that play has significant roles to ill children's treatment and recovery, we proposed and designed healing environment designs with functions of play at the hospital block, outpatient area and healing garden.

从成育庭园望向病房楼

成育庭園から病棟を望む

Viewing hospital building from 'Healing garden', seiiku large

屋顶庭园
屋上庭園
Roof-top garden

入口中庭
エントランスアトリウム
Entrance atrium

儿童大厅
チルドレンズモール
Children's mall

被绿色环抱的成育庭园
緑に包まれた成育庭園
'Healing garden' covered by full of green

平安医院

所 在 地　东京都练马区
竣工时间　1999年3月
建筑结构　钢筋混凝土结构,钢框架钢筋混凝土结构/地上3层，地下1层
用地面积　3150.02 m²
占地面积　1259 m²
建筑面积　3650 m²
获奖经历　1994年东京建筑奖

该建筑用地位于幽静的住宅区内，是一座拥有24间病房和84张床位的医院。用地的方向为东西向，中央部分设计了一个中庭，种植楠木。建筑整体空间呈回游性结构形态。

やすらぎミラージュ

所 在 地　東京都練馬区
竣 工 年　1999年3月
規　　模　RC造、SRC造／地上3階、地下1階
敷地面積　3,150.02 m²
建築面積　1,259 m²
延床面積　3,650 m²
受 賞 歴　1994.東京建築賞

敷地は閑静な住宅地にあり、東西に長い敷地の中央にクスノキのある中庭を設け、回遊性のある空間構成とした。収容規模は、24室・84床である。

Yasuragi Mirage

Location　Nerima-ku, Tokyo
Completion　March 1999
Structure　SRC, RC/1 floor below-grade, 3 above-ground floors
Site area　3,150.02 m²
Building area　1,259 m²
Total floor area　3,650 m²
Awards　1994 Tokyo Architecture Award

Sited in a quiet residential area, with its longitudinal axis from east to west, and centered on a courtyard with a kusunoki tree at its heart, the building has the purpose of creating a spatial structure with circulation. Capacity is 24 rooms and 84 beds.

从中庭望向建筑物

中庭から建物を望む
Viewing the building from the courtyard

公共房间

デイルーム
Communal room

浅井健康公园

所 在 地　滋贺县东浅井郡
竣工时间　2004 年 3 月
建筑结构　钢筋混凝土结构，钢结构 / 地上 2 层，地下 1 层
用地面积　20835 m²
占地面积　2819.27 m²
建筑面积　3546.98 m²

本建筑是在全国举行的老人诊所设计竞赛中被评选为最优秀的设计方案，它是由诊疗所、养护中心、温浴设施、餐厅、小规模高尔夫球场等健康体育设施所组成的综合健康设施。在设计上力求与周围环境相协调，采用传统的坡屋顶形式。为了方便当地村民放心利用，整体建筑除部分两层之外其余都是一层平房。

健康パークあざいプロジェクト

所 在 地　滋賀県東浅井郡
竣 工 年　2004 年 3 月
規　　模　RC 造、S 造／地上 2 階地下 1 階
敷地面積　20,835 m²
建築面積　2,819.27 m²
延床面積　3,546.98 m²

全国公募のプロポーザルによって最優秀案として選定された老人のための診療所、デイケアセンター、温浴施設、レストラン、グラウンドゴルフ等の健康スポーツ施設の合築した健康複合施設である。街並みにあわせた切妻屋根をもち、一部2階があるが、ほとんど平屋の町民が利用しやすく、また心安らぐ施設を目指している。

Azai Wellness Park Project

Location　Higashi Azai-gun, Shiga Prefecture
Completio　March 2004
Structure　RC and Steel /1 floor below-grade, 2 above-ground floors
Site area　20,835 m²
Building area　2,819.27 m²
Total floor area　3,546.98 m²

This multi-use healthcare facility containing a geriatric hospital, day care center, spa facility, restaurant, and such health sports facilities as ground golf, all constructed together, was selected as the most outstanding proposal in a nationwide public design competition. Its gabled roof blends in with the town; and, though some sections do have second floors, it is primarily ground floor-only, a facility that intends for town residents to have ease-of-use and a sense of comfort.

章佑会　关町之家

所 在 地	东京都练马区
竣工时间	2004 年 2 月
建筑结构	钢筋混凝土结构 / 地上 3 层
用地面积	2752.17 m^2
占地面积	1238.09 m^2
建筑面积	2748.75 m^2

本建筑是为智力障碍者所提供的住宅和工作设施。在住宅部分，将轻度障碍者和重度障碍者的活动领域分开布置，二者之间通过中庭连接起来。在室内设计方面力图营造出开敞、欢快的氛围，同时也试图为患者们提供一个可以增进交流的场所。

章佑会　関町ホームプロジェクト

所 在 地	東京都練馬区
竣 工 年	2004 年 2 月
規　　模	RC 造／地上 3 階
敷地面積	2,752.17 m^2
建築面積	1,238.09 m^2
延床面積	2,748.75 m^2

知的障害者のための住居と作業施設である。住居部分は比較的軽い障害者の階と重い障害者の階を分けながら、中庭を介して相互の存在を認識できる空間構成がとられている。全体に明るさと楽しさがあり、交流が促進されるようなインテリアが目指されている。

Sekimachi Nursing Home Project

Location	Nerima-ku, Tokyo
Completion	February 2004
Structure	RC/3 above-ground floors
Site area	2,752.17 m^2
Building area	1,238.09 m^2
Total floor area	2,748.75 m^2

A residence and work facility for persons with intellectual disabilities. Although the residence portion segregates the floors for people with rather light disabilities from the floors for those with severe disabilities, it possesses a spatial structure in which mutual existence is recognizable across the courtyard and an interior with an overall bright and fun feeling to promote intercommunication.

Urban

城市设计

アーバンデザイン

URBAN

85' 筑波科技博览会儿童广场

所 在 地　　茨城县筑波市
竣工时间　　1985 年 3 月
建筑结构　　30000 m^2

该儿童广场是由筑波科学万国博览会为政府提供的展示设施，位于博览会场的中央部位。整个广场被划分为四个主体空间，其一是被称为"不可思议的庭院"的入口广场。利用错视效果营造了室外版诶姆兹的房间，设计了不可思议的回廊、门的迷宫等装置，是可以在这里体验到不可思议的小宇宙奥秘的广场。其二是连续的全长 270m 被称为是"趣味筒管"的科学体验游戏玩具。其三是由动物信号、机械汽车、震级地板（地球体验装置）构成的"地球和机械动物园的广场"。其四是"日本列岛"，将 10 万分之一的地图作为直径 36m 的地球的一部分表现出来。

科学万博　つくば '85 こども広場

所 在 地　　茨城県つくば市(筑波研究学園都市)
竣 工 年　　1985 年 3 月
敷地面積　　30,000 m^2

筑波科学万国博覧会政府出展施設であるこども広場は、博覧会場中央部分に位置した。四つのテーマ空間に分かれ、第一は「ふしぎの庭」と呼ぶエントランス広場である。錯視を利用した屋外版エイムズの部屋ともいうべきふしぎ回廊、ドアの迷宮など、不思議な小宇宙が体験できる広場である。第二は「おもしろチューブ」と呼ぶ、全長270mの連続科学体験遊具。第三はアニマルアイズ、メカ車、マグニチュードの床（地球体験装置）で構成される「地球とメカ動物園の広場」、第四は「日本列島ゾーン」で10万分の1の地図を直径36mの地球の一部として表現した。

Children's Plaza, Tsukuba Expo '85

Location　　Tsukuba City, Ibaraki Prefecture
Completion　　March 1985
Site Area　　30,000 m^2

This children's plaza, a facility exhibited at the Tsukuba Science Expo, is situated at the center of the Expo site. It is divided into four themes. The first is "Garden of Wonder", which features optical illusions. The second is a "Fun Tube", which is a 270-m-long, 2.7-m-diameter Tube Structure serving as a science exploratorium. The third is a "Mechanimal", and the fourth is the "Japanese Archipelago Zone", which re-creates a sector of the globe at a scale of 1/100,000, stretching 36 m across.

管状的巨大游具"有趣的管道"
チューブの巨大遊具'おもしろチューブ'
Giant tube play structure 'Fun tube'

水泵站
ポンプステーション
Pump station

FUJITSU
FUYO

帐篷的巨大游具“山脉滑行者”

テントの巨大遊具‘マウンテンスライダ

Giant tent play structure 'Mountain slider'

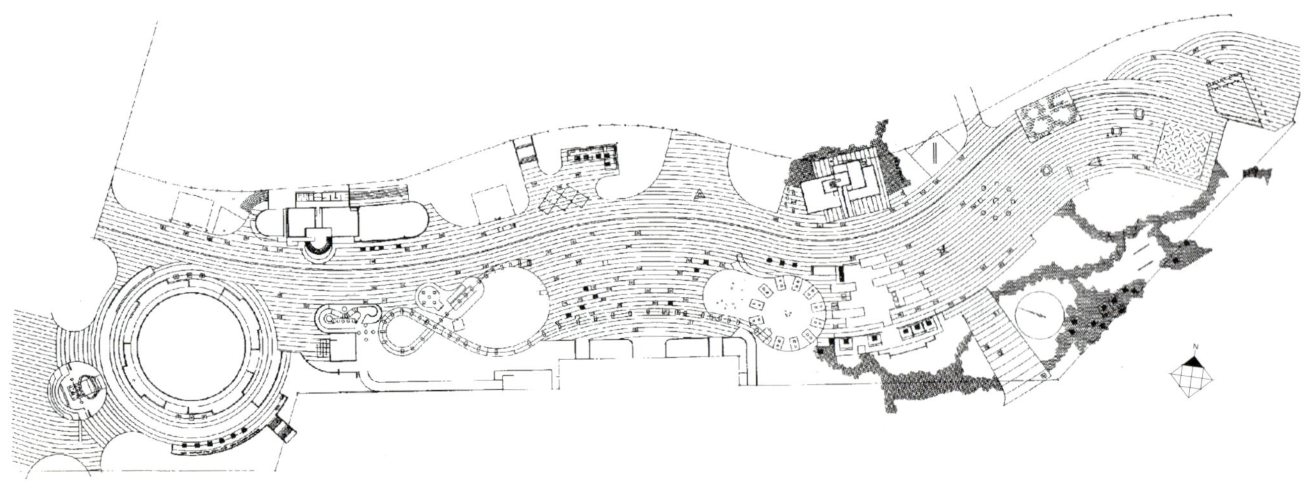

日本列岛区
日本列島ゾーン
Japanese archipelago zone

奇怪的画
ふしぎ絵
Strange picture

扭曲的房间
歪んだ部屋
Crooked space

TSUKUBA EXPO'85 TIMESPACE

藤野艺术之家

所 在 地　神奈川县津久井郡
竣工时间　1995年8月
建筑结构　钢筋混凝土结构，部分为钢框架钢筋混凝土结构／地上5层
用地面积　26160.83 m^2
占地面积　3257.67 m^2
建筑面积　5422.25 m^2
获奖经历　1996年商业环境设计奖／神奈川建筑竞赛优秀奖

藤野町以战争年代东京的艺术家们疏散到这里，并形成艺术村而闻名，即使是现在，依然有50多位艺术家居住在这里。人们大多是因为艺术才到这里的，因此藤野艺术之家的设计应该作为町的标志性设施来进行。该设施为人们所知的最大原因是人们在这里住宿的同时，还可以通过进行各种各样的艺术体验和自然体验，培养丰富的感性和创造性，所以成为人们所喜爱的宿泊设施。

艺术楼、工作室

芸術棟、工房
Art Block, Workshop room

藤野芸術の家

所 在 地　神奈川県津久井郡
竣 工 年　1995年8月
規　　模　RC造一部SRC造／地上5階
敷地面積　26,160.83 m^2
建築面積　3,257.67 m^2
延床面積　5,422.25 m^2
受 賞 歴　1996. 商環境デザイン賞、神奈川建築コンクール優秀賞

藤野町は、戦時中東京の芸術家達が疎開し、芸術村をつくったことで有名で、現在も50人の芸術家達が在住している。芸術による町おこしを行っているが、藤野芸術の家はそのシンボル施設として計画された。宿泊しながら様々な芸術体験や自然体験を通して、豊かな感性と創造性を育める施設として知られ、若いファミリーから熟年グループまで、大人気の施設となっている。

Fujino Workshop for Art

Location	Tsukui-gun, Kanagawa Prefecture
Completion	August 1995
Structure	RC, part SRC/5 above-ground floors
Site area	26,160.83 m^2
Building area	3,257.67 m^2
Total floor area	5,422.25 m^2
Awards	1996 JCD Award, Kanagawa Architecture Conquer Superior Award

Fujino Town is famous for being an artists' colony created by artists who evacuated Tokyo during wartime, and at present 50 artists are in residence. The town had been developed economically through art and the Fujino Workshop for Art was planned as a facility symbolic of this. Known as a facility in which profuse sensitivity and creativeness can be nurtured during stays through the experience of art and nature, it is very popular with all, from young families to groups of senior citizens.

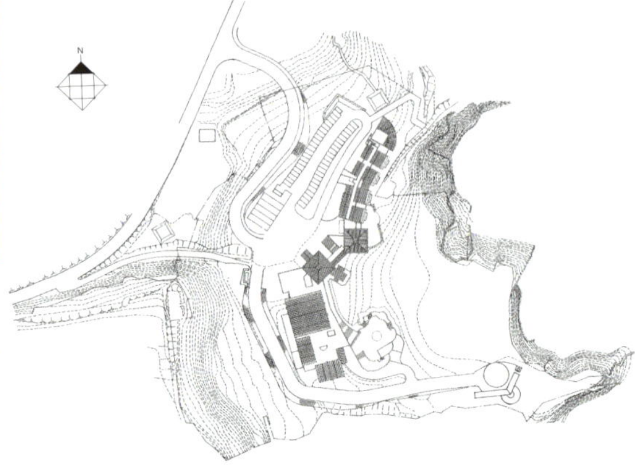

空中摄影，右侧是宿泊楼，左侧是艺术楼

空撮. 右側が宿泊棟、左側は芸術棟

Aerial view. Lodging block on the right and workshop block on the left

设置有主入口的西侧外观

メインエントランスのある西側外観

External view of the building with the entrance from the west

外墙材料使用硅藻土
外壁は珪藻土を使用
Diatomaceous earth is used for the external wall

东侧外观，面前是宽敞的草坪广场
東側外観.手前に芝生広場が広がる
External view from the east. There is a grass open field on the front

连接艺术楼与住宿楼的桥

芸術棟と宿泊棟を結ぶブリッジ

Bridge connecting Art Block and Lodging Block

艺术楼、入口大厅

芸術棟、エントランスホール

Art Block, entrance hall

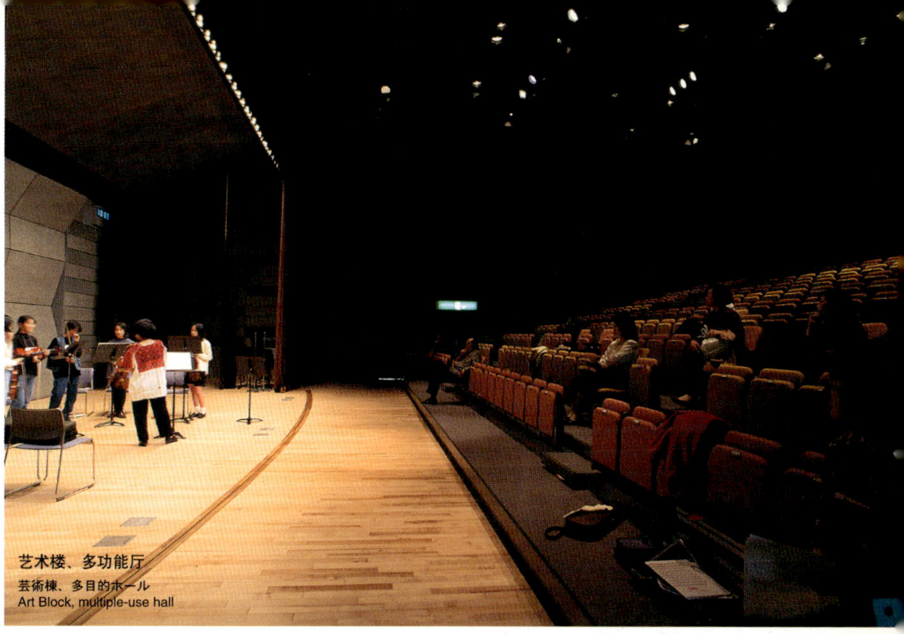

艺术楼、多功能厅

芸術棟、多目的ホール

Art Block, multiple-use hall

住宿楼，餐厅区域

宿泊棟、レストランエリア

Lodging block, restaurant area

富岩运河环水公园

所 在 地	富山县富山市
竣工时间	1999年11月
建筑结构	钢框架钢筋混凝土结构 / 地上3层（展望台），钢筋混凝土结构 / 地上1层（泉水瀑布广场）
用地面积	97000 m²
占地面积	121.1 m²（展望台），775.8 m²（泉水瀑布广场）
建筑面积	289.4 m²（展望台），775.8 m²（泉水瀑布广场）
获奖经历	2000年城市景观大奖地区级部门，2001年全建奖城市部门，优良设计奖、城市环境设计会议2000JUDI奖优秀奖、日本造园学会作品选集

以富岩运河为中心进行车站周围地区的再开发计划，本设计就是在1989年富山县运河公园指名设计竞赛中作为最优秀设计方案而被选定为实施方案的。现在，在环水公园的旁边，广场的对面建成了县级设施以及市体育馆，希望借此能够吸引更多的市民来利用它。

富岩運河環水公園

所 在 地	富山県富山市
竣 工 年	1999年11月
規　　模	SRC造／地上3階（展望塔）、RC造／地上1階（泉と滝の広場）
敷地面積	97,000 m²
建築面積	121.1 m²（展望塔）、775.8 m²（泉と滝の広場）
延床面積	289.4 m²（展望塔）、775.8 m²（泉と滝の広場）
受 賞 歴	2000.都市景観大賞地区レベル部門、2001.全建賞都市部門、グッドデザイン賞、都市環境デザイン会議2000、JUDI賞優秀賞、日本造園学会作品選集

本プロジェクトは、富岩運河を中心として駅周辺を再開発するということで、1989年富山県カナルパーク指名設計競技によって最優秀案として選定された。富岩運河の最終的なエッジは、直径35 mの水盤によってイベンタブルな空間としてよみがえった。現在、環水公園に隣接し、広場を挟んで県立施設、市立体育館ができ、多くの市民の利用が図られている。

Fugan Canal Park

Location	Toyama City, Toyama Prefecture
Completion	November 1999
Structure	SRC/3 above-ground floors (Tower); RC/1 above-ground floor (Fountain Square)
Site area	97,000 m²
Building area	121.1 m²(Tower); 775.8 m²(Fountain Square)
Total floor area	289.4 m²(Tower); 775.8 m² (Fountain Square)
Awards	2000 Urban Landscape Award; 2001 Zenken Award (Urban Division), Good Design Award, JUDI Award, Works of Japanese Institute of Land scape Architecture

A project for the redevelopment of the station area centering on Fugan Canal, and selected as the most outstanding proposal in the Toyama Prefecture Canal Park Designated Design Competition of 1989. The final edge of Fugan Canal was revived as an event worthy space with a flower ball of 35m in diameter. At present, prefectural facilities and municipal gymnasium have been built across the square next to the Kansui Park for use by numerous residents.

长58m的天门桥黄昏景色
長さ58mの天門橋の夕景
View of Tenmon Bridge (58m length) at dusk

正面回廊的圆形水盘和呈幕状落下的瀑布

フロントデッキの円形水盤とカーテン状に落ちる滝

Round pond at front deck area, and waterfall flowing like curtain

从观览塔眺望正面回廊

展望塔からフロントデッキをのぞむ

The front deck seen from the observatory tower

举办晚会的场所－山手游艇俱乐部

所 在 地　神奈川县横滨市
竣工时间　1987 年 4 月
建筑结构　钢结构 / 地上 3 层
用地面积　902.48 m^2
占地面积　226.63 m^2
建筑面积　354.08 m^2
获奖经历　1988 年商业环境设计奖

在法律上商店的规模被限制在$50m^2$以下。本设计将以贩卖三明治为主的平房咖啡店和三层的住宅楼布置成L形，并在道路一侧设置停车场。因为用地的南侧地势较低，所以从道路一侧看住宅部分只能看到二层。在建筑物的周围种植的大多都是美国四照花和光叶榉树，每到4、5月份花开的时候非常美丽。

パーティハウス・山手ヨットクラブ

所 在 地　神奈川県横浜市
竣 工 年　1987 年 4 月
規　　模　S 造／地上 3 階
敷地面積　902.48 m^2
建築面積　226.63 m^2
延床面積　354.08 m^2
受　　賞　1988. 商環境デザイン賞

山手は高級住宅地であるが、町の性質上店舗は 50 m^2 以下に法的に制限されている。サンドイッチをメインとする平屋のコーヒーショップと3階建ての住居棟をL型に構成し、道路側に駐車のスペースをとっている。敷地が南側に下がる形になっているため道路側からは住居部分は2層に見える。この建物が最も美しく見えるのは、建物の周囲に植栽されたハナミズキが咲く4月～5月にかけてである。

Party House - Yamate Yacht Club

Location　Yokohama City, Kanagawa Prefecture
Completion　April 1987
Structure　Steel/3 above-ground floors
Site area　902.48m^2
Building area　226.63m^2
Total floor area　354.08m^2
Awards　1988 JCD Award

Yamate, a high–class residential area, has been limited by code to a maximum of $50m^2$. The one-story coffee shop is parallel to the street with the three-story residences behind it running perpendicular to it. Since the site slopes to the south, the residential part appears to be two-storied from the street. The project looks best in April and May, when the dogwood blooms planted around the building.

中庭内的泳池
中庭プール
Swimming pool in the garden

从室内看游泳池

室内からプールを望む

Viewing the pool from the inside

玄关

玄関

Entrance hall

起居室

居間

Living room

举办晚会的场所－山手游艇俱乐部
パーティハウス・山手ヨットクラブ
Party House - Yamate Yacht Club

宇宙阿伊卢羽咋

所 在 地	石川县羽咋市
竣工时间	1996年3月
建筑结构	钢筋混凝土结构，部分为钢结构·钢框架钢筋混凝土结构/地上4层，地下1层
用地面积	31104 m²
占地面积	4847 m²
建筑面积	6750 m²
获奖经历	1997年石川县建筑奖石川县知事奖

石川县羽咋市因其市内的气多神社所流传的与UFO有关的传说而远近闻名，作为城市建设的标志物而建造的本建筑是由宇宙展示室、图书馆、多功能音乐厅、终生学习等组成的综合文化设施。环绕着洛彤达的是带有顶层天窗的三层吹拔走廊，是连接各种综合功能的多功能回廊空间。

コスモアイル羽咋

所 在 地	石川県羽咋市
竣工年月	1996年3月
規　　模	RC造一部S造・SRC造／地下1階・地上4階
敷地面積	31,104 m²
建築面積	4,847 m²
延床面積	6,750 m²
受 賞 歴	1997.石川県建築賞石川県知事賞

石川県羽咋市は気多神社に伝わる伝説に関連したUFOによる町おこしが有名で、その町づくりのシンボルとして宇宙展示室、図書館、多目的音楽ホール、生涯学習からなる文化複合施設がつくられた。ロトンダを取り囲むトップライトを持つ3層吹き抜けのギャラリーは、複合的な機能を結びつける多目的な廊空間となっている。

Cosmo Isle Hakui

Location	Hakui City, Ishikawa Prefecture
Completion	March 1996
Structure	RC, part Steel, SRC/1 floor below-grade, 4 above-ground floors
Site area	31,104 m²
Building area	4,847 m²
Total floor area	6,750 m²
Awards	1997 Ishikawa ArchitectureAward the Governor Award

Hakui City, Ishikawa Prefecture is famous for the town's economic development in by UFO in relation to the legend of Keta Shrine. As a symbol for this purpose, a cultural multi-facility consisting of a space exhibition room, library, multi-purpose music hall and lifetime study area was created. A three-story open gallery with skylights encircling the rotunda functions as a multipurpose corridor space linking up the multiple functions.

图书馆区域
図書館エリア
Library space

UFO 大厅
大ホール「ユーフォニー」
Large hall, "UFONIE"

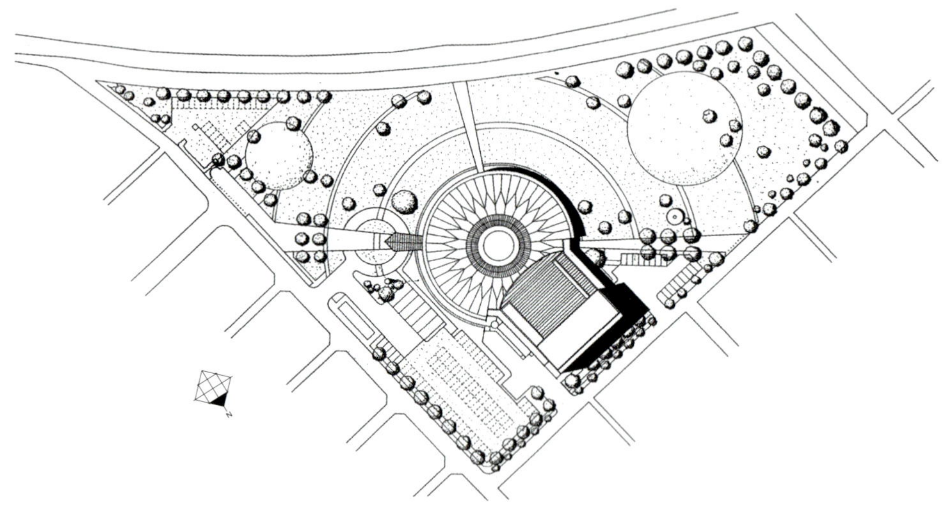

以宇宙飞船为意象的外观和火箭

宇宙船をイメージした外観とロケット

External view of the building with image of UFO and a rocket

入口大厅
エントランスロビー
Entrance lobby

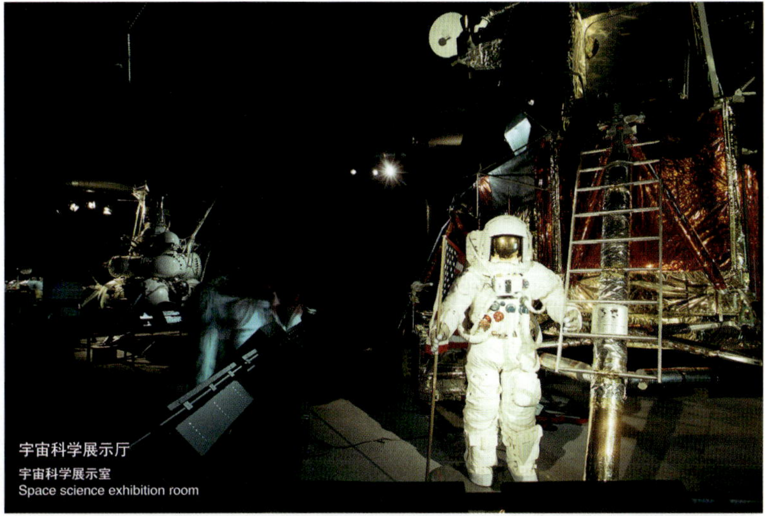

宇宙科学展示厅

宇宙科学展示室

Space science exhibition room

圆形的多目的厅“洛形达”

円形の多目的な小ホール「ロトンダ」

Small round multiple-use hall, "ROTONDA"

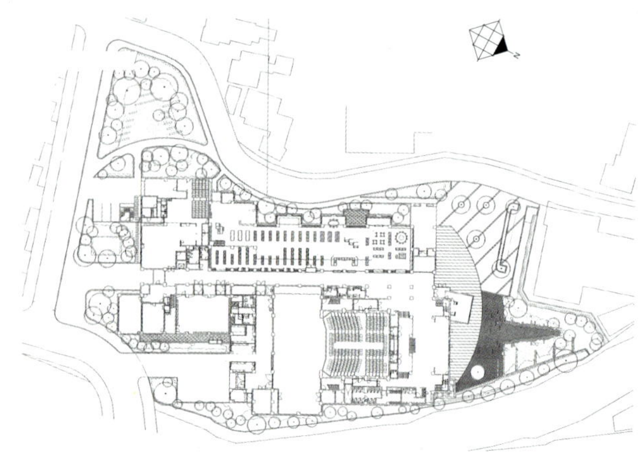

城岛町综合文化中心

所 在 地　福冈县三潴郡
竣工时间　1998年3月
建筑结构　钢筋混凝土结构 / 地上3层
用地面积　11959 m²
占地面积　4124 m²
建筑面积　5965 m²

位于福冈县南部的城岛町曾经以造酒和生产瓦而繁荣一时，因此酒窖这种能够勾起人们乡愁的风景就成为本建筑的设计原点。相对于街道尺度而显得体量过大的会馆，具有连续酒窖形屋顶的图书馆，意在创造出以书为核心的，并能促进人与人之间相互沟通交流的空间。文化中心的另一个核心是印伽托会馆的观众席，它再现了曾经是城市文化·娱乐殿堂的城岛剧场的氛围，是和洋折衷的华丽空间，表现了当时充满活力的大众文化的兴隆昌盛。

城島町総合文化センター

所 在 地　福岡県三潴郡
竣工年月　1998年3月
規　　模　RC造／地上3階
敷地面積　11,959 m²
建築面積　4,124 m²
延床面積　5,965 m²

福岡県南部に位置する城島町はかつて酒造りと瓦の生産で栄え、酒蔵という郷愁を誘う風景が本施設の原点となっている。町並みに対して大きすぎるホールのスケールを、連続する蔵形の屋根で緩和した図書館は、書物を核に人と人が出会い、育つことを大切にした空間づくりを目指している。文化センターの一方の核であるインガットホールの客席は、かつて町の文化・娯楽の殿堂であった城島劇場の雰囲気を再現した和洋折衷の晴れやかな空間で、筑豊・筑後が活気に満ちていた時代の大衆文化の興隆を表現している。

Jojima Cultural Center

Location　Mizuma-gun, Fukuoka Prefecture
Completion　March 1998
Structure　RC/3 above-ground floors
Site area　11,959 m²
Building area　4,124 m²
Total floor area　5,965 m²

Jojima Town, located in the southern part of Fukuoka Prefecture, previously flourished with Sake making and production of roofing tiles. Nostalgic scenes of sake storehouses are the source of this facility. The library, which moderates the overly large scale of the hall for the town by means of serial storehouse-type roofs, is aimed at creating a space for people to meet each other and develop themselves with a focus on books. Guest seating at Ingot Hall, which is the other center of Culture Center, makes up a bright space with a mixture of Japanese and Western styles, representing the atmosphere of Jojima Theater, previously the pantheon of culture and entertainment for the town, and express the flourishing of mass culture in the age when the atmosphere in the Chikuho and Chikugo areas was brimming with activity.

拥有600个观众席的印伽托会馆
客席数６００のインガットホール
'Ingot Hall' with 600 seats

连接大厅和图书馆的画廊
ホールと図書館をつなぐギャラリー
Gallery space connecting the hall and library

图书馆
図書館
Library

图书馆
図書館
Library

世界淡水鱼园(绿洲公园)

所 在 地	岐阜县羽岛郡
竣工时间	1999 年 4 月
建筑结构	钢结构平房，(拱廊)地上 2 层
用地面积	34000 m²
占地面积	419.29m²(绿洲咖啡厅,绿洲商场)，351.01m²(岐阜之最)，314.04 m²(濑贺世界)，399.48 m²(拱廊·大屋顶)
建筑面积	384.48 m²(绿洲咖啡厅,绿洲商场)，307.25 m²(岐阜之最)，286.59 m²(濑贺世界)，555.13 m²(拱廊·大屋顶)
获奖经历	2000 年商业环境设计奖，2002 年 American Wood Design 奖

绿洲公园是国营木曾三川公园的一部分，由河川环境乐园和高速道路川岛服务区以及高速道路绿洲所组成的新型综合设施。3ha 的用地。第 1 期工程为商业设施，第 2、3 期为水族馆工程，其中第一期工程的商业设施已经完工，包括贩卖、娱乐、虚拟餐厅等，它不仅营造出水边空间的热闹氛围，而且商业活动也非常成功。

世界淡水魚園「オアシスパーク」

所 在 地	岐阜県羽島郡
竣工年月	1999 年 4 月
規　　模	S造　平屋、(ガレリア・フィッシュオンチップス) 地上 2 階
敷地面積	34,000 m²
建築面積	419.29m² (オアシスカフェ、オアシスマート) 351.01m² (ギフベスト) 314.04m² (セガワールド) 399.48m² (ガレリア・大屋根)
延床面積	384.48m² (オアシスカフェ、オアシスマート) 307.25m² (ギフベスト) 286.59m² (セガワールド) 555.13m² (ガレリア・大屋根)
受 賞 歴	2000.商環境デザイン賞　2002.アメリカンウッドデザインアワード

国営木曽三川公園の一部、河川環境楽園と高速道路川島サービスエリア及び、ハイウエイオアシスと複合した新しいタイプの官民共同プロジェクトである。敷地面積は3ha、第一期の商業エリアと第二、三期の水族館エリアに分かれており、第一期商業エリアが完成した。商業施設エリアは、物販、ゲーム、バーチャルレストラン等からなっており、水辺空間のにぎわいを作りだし、商業的にも成功している。

'Oasis Park'

Location	Hashima-gun, Gifu Prefecture
Completion	April 1999
Structure	Steel/2 above-ground floors (Galleria, Fish on Chips)
Site area	34,000 m²
Building area	419.29 m² (Oasis Café, Oasis Mart); 351.01 m²(Gifu Best); 314.04 m² (Sega World); 399.48 m² (Galleria, Roof Square)
Total floor area	384.48 m²(Oasis Café, Oasis Mart); 307.25 m² (Gifu Best); 286.59 m²(Sega World); 555.13 m² (Galleria, Roof Square)
Awards	2000 JCD Award; 2002 American Wood Design Awards

This is new type joint public and private project in which Kasen Kankyo Rakuen, a part of Kiso Sansen National Park and Kawashima Highway Service Area and highway rest stop are merged. The site is on 3ha, and is divided into a first stage commercial area and second and third stage aquarium areas. The commercial facilities area, composed of merchandising, gaming and virtual restaurant areas, etc., and with a lively waterfront space, has been a commercial success.

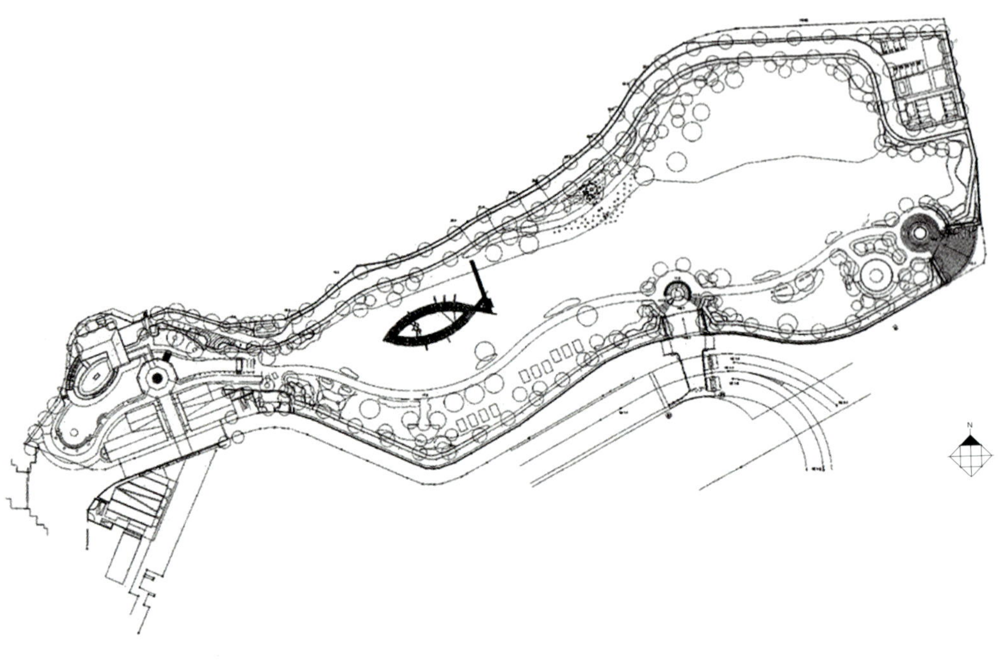

周长约 80m 的带有屋顶的鱼形状游戏玩具
魚の形をモチーフにした全周約 80 mの屋根付き遊具
Roofed play structure using a motif of shape of fish. The total length is 80 m

水路的右侧是"钩上之鱼"，左侧是"岐阜之最"

水路を挟んで右側が「フィッシュオンチップス」左側が「ギフベスト」

"Fish on Chips" on the right and "Gifu Best" on the left over the pond

拱廊的正面

正面は「ガレリア」

"Galleria" on the front

餐厅
レストラン
Restaurant

拱廊内部的喷泉
ガレリア内部の噴水
Fountain in the 'Galleria'

横须贺市三笠公园大街

所 在 地	神奈川县横须贺市
竣工时间	1987 年 3 月
建筑结构	全长约 480m
占地面积	10000 m^2
获奖经历	1990 年日本建筑学会作品选集

该三笠公园大街是沿横须贺海 1 万 m 散步路的模范地段，是从美军基地入口开始到海边三笠公园的约 300m 的路段。在车道的一侧设计了约 26m 的绿化带，并将该绿化带划分为以水之广场、柱之广场、风之广场和戏水广场为核心的四个区域，并通过水系将它们连接成为一个整体。

横須賀市三笠公園通り

所 在 地	神奈川県横須賀市
竣 工 年	1987 年 3 月
全 長 約	480 m
敷地面積	10,000m^2
受 賞 歴	1990.日本建築学会作品選集

この三笠公園通りは横須賀海沿いの 1 万 m プロムナードのモデル区域で、米軍基地の入り口から海側の三笠公園まで約 300m の距離をもつものである。車道を片側に寄せ、約26m緑道をつくることを提案した。この緑道は4つのブロックに分けられ、水の広場、マストの広場、ジャブジャブ広場、風の広場という4つの核をもっており、これらは水の小さな流れによって連結している。

Yokosuka Mikasa Park Street

Location	Yokosuka City, Kanagawa Prefecture
Completion	March 1987
Length	480m
Site area	10,000 m^2
Awards	1990 Architectural Institute of Japan Selected Architectural Designs

This Mikasa Park Street, which has 300 meter long between the American Base Camp and the Mikasa Park, is at the ten thousand meters promenade's midsection along the seaside of Yokosuka and stands as a symbol of the whole. One side of road from Mikasa Park is nearly 26m wide, of which roadway is given over to pedestrian uses and arranged around four plazas: the Water, Mast, Wind and Splash plazas. These plazas are connected with a water stream.

公园大道上看海
公園通りから海を望む
Viewing the sea from the park street

桅杆广场
マストの広場
Plaza of mast

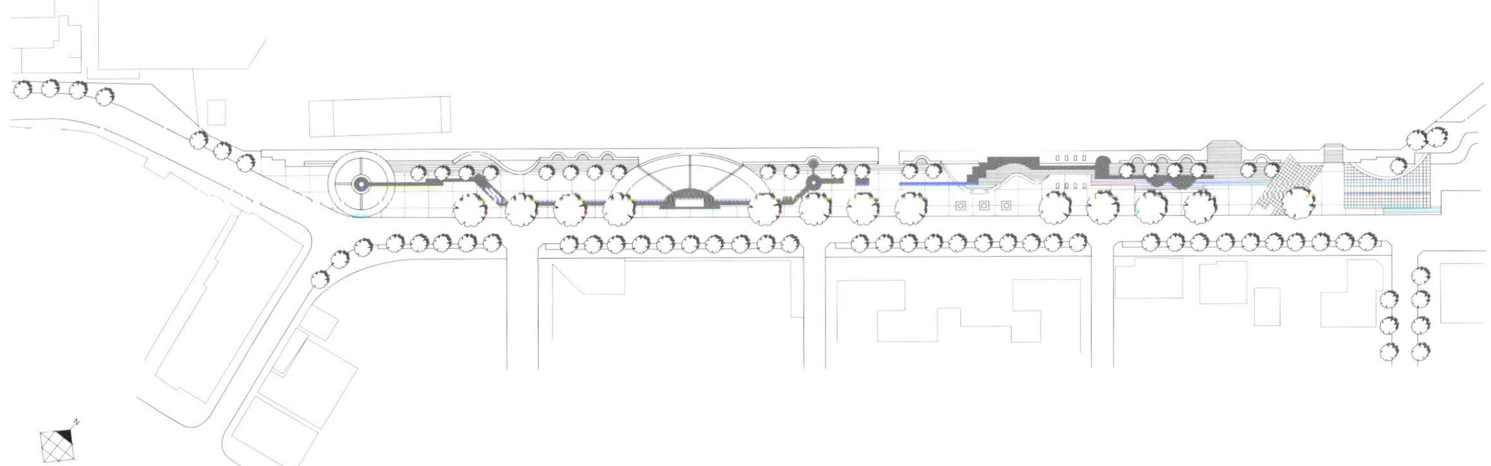

弁天町水之乐园(游泳池)

所 在 地	大阪府大阪市港区
竣工时间	1990 年 7 月
建筑结构	规模 钢框架钢筋混凝土结构 / 地上 5 层，地下 3 层，屋顶间 1 层
用地面积	30123.48 m²
建筑面积	11160.37 m²(仅游泳池)
获奖经历	1991 年商业环境设计奖

本建筑是位于大阪市弁天町再开发地区一角的立体娱乐设施，在平面长 150m，宽 75m 范围的室内空间拥有目前国内最大规模的水上乐园。室内地势中央低，两端高，呈丘陵状的起伏，通过称为天河的水桥来联系它们之间的交通。

弁天町ウォーターランド「プールズ」

所 在 地	大阪府大阪市港区
竣 工 年	1990 年 7 月
規　　模	SRC 造／地上 5 階、地下 3 階、塔屋 1 階
敷地面積	30,123.48m²
延床面積	11,160.37m²（プールズのみ）
受 賞 歴	1991.商環境デザイン賞

大阪市弁天町の再開発ブロックの一角に立体的アミューズメント施設として計画された。平面的に長さ 150 m、巾 75 mの広さで、屋内では国内最大規模の水のプレイランドである。中央部が低く、両端部が丘状となっており、それをスカイリバーという水の橋がつないでいる。

Bentencho Water Land 'Pools'

Location	Minato-ku, Osaka City
Completion	July 1990
Structure	SRC/3 floor below-grade, 5 above-ground floors
Site area	30,123.48m²
Total floor area	11,160.37m² (Pools)
Awards	1991 JCD Award

This amusement facility built as part of the development taking place in the Bentencho district of Osaka, is 150 meters long and 75 meters wide, making it the largest indoor water amusement facility in Japan. The central area is lower than the two ends which are terraced and they are connected by what is called the "Sky River".

高 5m 的天河使游泳的人们感觉如同浮游在空中一样

高さ 5 mのスカイリバーは空中浮遊感覚を泳ぐ人に与える

The 5m high "Sky River" elevated channel gives bathers a chance to experience a sense of playing in mid-air

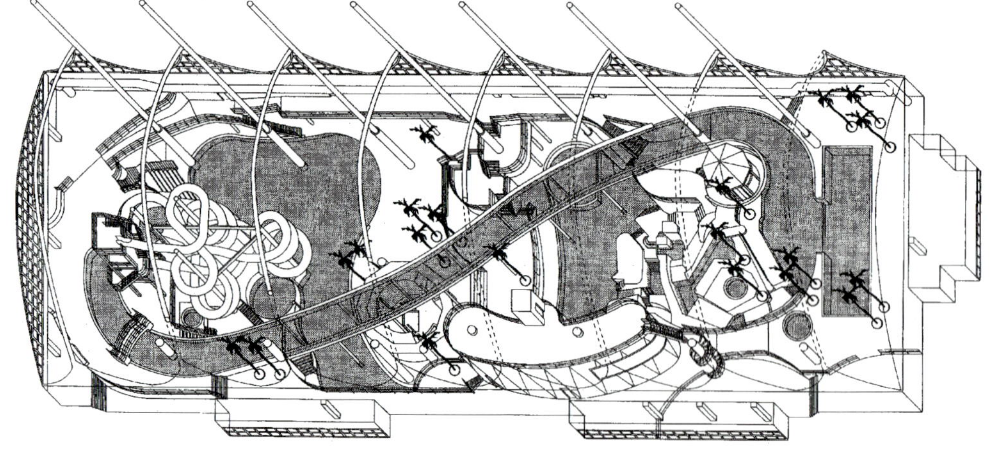

各种水池标高不同，形成丰富的立体形态

各種プールのレベルに変化をもたせ、立体的に構成している

A three-dimensional composition has been achieved with the changing levels of each pool

营团地铁南北线

所 在 地	南北线驹入站 - 赤羽岩渊站之间
竣工时间	1991 年 11 月
用地面积	地面建筑50 m²-100 m²,中央大厅500 m²-1000 m² × 6站，站台 600 m² × 6 站
获奖经历	1992年商业环境设计奖，展示设计奖

该地铁站首次在日本采用了站台门的设计理念。考虑到因此而对于站台上候车的人们所产生的闭塞感，所以在站台的对面设计了艺术墙。连续且色彩丰富的艺术墙还可以布置企业的商业广告。车站地面部分的屋顶以列车的集电弓架为主题，设计成悬浮在空中的屋顶形式。

地铁站台座椅
地下鉄ホームの座席
Hall seats in subway station

営団地下鉄南北線

所 在 地	南北線駒込駅 - 赤羽岩淵駅間
竣 工 年	1991 年 11 月
敷地面積	上屋 50-100m² コンコース 500-1,000m² × 6 駅 ホーム 600m² × 6 駅
受 賞 歴	1992. 商環境デザイン賞、ディスプレイデザイン賞

日本ではじめて地下鉄にホームドアを採用した。そのため、ホームの人々にとって閉塞感を増す心配がもたれ、私達はホームの対向面にアートウォールというシステムを提案した。色彩豊かな連続した芸術的な壁面をつくり、企業にスポンサーになってもらうというシステムである。地下鉄上屋についてはパンタグラフをモチーフにして空中に浮かぶ屋根のデザインを行った。

Eidan Subway, Nanboku Line

Location	Nanboku Line between Komagome Station and Akabane Iwabuchi Station
Completion	November 1991
Site Area	50-100m² (Station Roof); 500-1,000m² × 6 (Concourse); 600m² × 6 stations (Platform)
Awards	1992 JCD Award, Display Design Award

Platform doors were adopted for the very first time in Japan for this scheme. However, because the use of such doors creates a rather claustrophobic atmosphere on the platforms, we proposed a scheme for decorating the tunnel walls beyond the doors to try and alleviate this. Instead of having an array of commercial advertising, companies were asked to sponsor the production of colorful murals. For the station roof, a suspended train pantograph motif was used.

出入口建筑，大量地使用玻璃试图使设计显得轻快活泼
出入口上屋 ガラスを多用した軽快なデザインを意図した
The use glass on the roof of the stairs going up to the exit is designed to create a light and cheerful atmosphere

屋顶采用了列车的集电弓架的形式
パンタグラフを利用した屋根
The roof with its pantograph motif

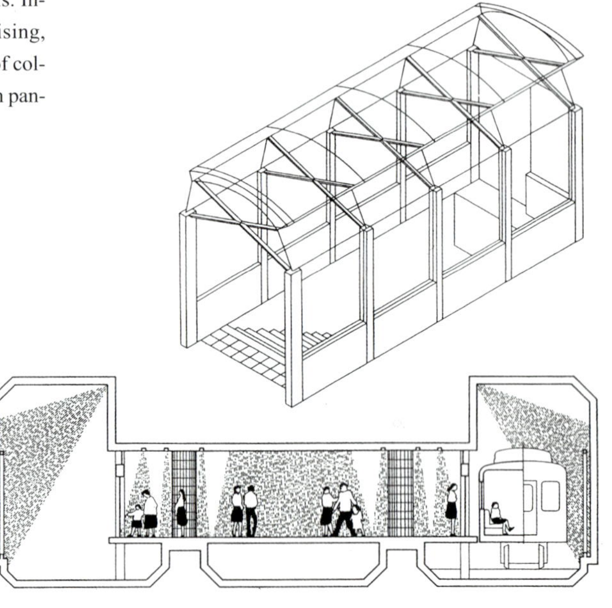

站台门和艺术墙，狭窄的地下空间因明亮的艺术大厅而发生变化

ホームドアとアートウォール.狭い地下空間を明るく楽しいアートホールによって変化させる

Doors the platform and decorated wall. The different motifs on the walls of each station brighten up the cramped space of the subway.

五藤光学研究所山梨工厂

所 在 地	山梨县中巨摩郡
竣工时间	1996 年 12 月
建筑结构	钢结构 / 地上 2 层
用地面积	14373 m²
占地面积	3352 m²
建筑面积	4355 m²
获奖经历	1997 年优良设计奖、SDA 奖

建筑内部主要分为两大部分，一部分是由入口·食堂·更衣室等所组成的区域，这部分区域的设计力求与外部优美的自然景观相互呼应，形成开放明亮的舒适空间。另一部分则是由监理·设计·开发·研究部门等所组成的区域，根据功能需要自然光很难进入到这部分空间。建筑的室外是铺有草坪的开放空间，同时也可以作为休息空间来使用。

五藤光学研究所山梨工場

所 在 地	山梨県中巨摩郡
竣工年月	1996 年 12 月
規　模	S 造／地上 2 階
敷地面積	14,373m²
建築面積	3,352m²
延床面積	4,355m²
受 賞 歴	1997. グッドデザイン賞、SDA 賞

施設は大きく2つのゾーンに分けられ、エントランス・食堂・更衣室は外部の恵まれた自然に対して、開放的で明るいくつろげる空間とした。監理・設計・開発・研究部門は、機能的欲求から自然光が入りにくい構造とした。外部には芝生のオープンスペースが設けられ、レクレーションスペースとして利用されている。

Goto Optical MFG. Yamanashi Factory

Location	Nakakoma-gun, Yamanashi Prefecture
Completion	December 1996
Structure	Steel/2 above-ground floors
Site area	14,373m²
Building area	3,352m²
Total floor area	4,355m²
Awards	1997 Good Design Award, SDA Award

This facility is divided into two primary zones, with the entrance, dining room and changing room created as open, bright and relaxing spaces oriented to the luxuriant nature outside. The control, design, development and research section, due to functional requirements, is created as structure into which natural light does not readily penetrate. There is an open lawn space created outside that is used as a recreation space.

员工餐厅
社員食堂
Cafeteria

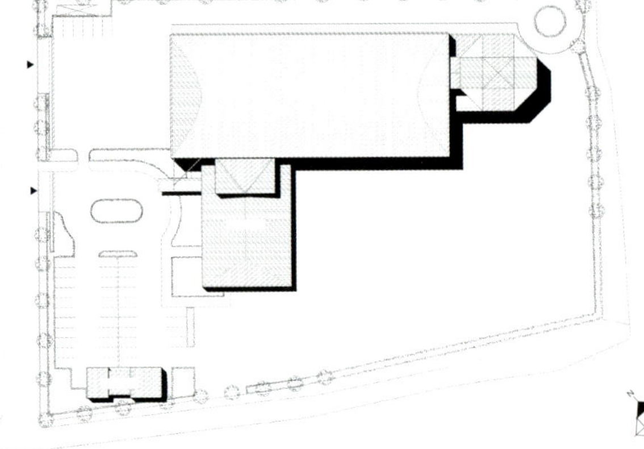

福利栋和生产栋，实验馆环绕草坪广场呈 L 型布置

厚生棟と生産棟、そして実験ドームが芝生広場を囲んで L 型に配置されている

Health and welfare block, production block, and experiment block are located in L-shape around grass square

大厅内览

ロビー内観

Interior of lobby

实验馆东侧外观

東側外観.実験ドーム

Eastern elevation. Experiment dome

和歌山县动物保护中心

所 在 地	和歌山县海草郡
竣工时间	2000 年 3 月
建筑结构	钢结构，部分为钢筋混凝土结构/地上 2 层，地下 1 层
用地面积	31750 m^2
占地面积	4255 m^2
建筑面积	4057.32 m^2

本动物保护中心是为了普及开发与动物接触时所需要的相关知识以及传授饲养动物的方法而建立的专业设施。在动物监理方面采用全部个别犬房的体制，这在全国也是最先行的尝试，因此可以说，在这里能够彻底学习到爱护动物的精神。

和歌山県動物愛護センター

所 在 地	和歌山県海草郡
竣工年月	2000 年 3 月
規　　模	S・RC 造　地下 1 階・地上 2 階
敷地面積	31,750m^2
建築面積	4,255m^2
延床面積	4,057.32m^2

動物に接するのに必要な知識と飼い方の普及啓発をするための施設として計画された。監理動物に対する個別犬房の全面的採用も全国ではじめての試みであり，動物愛護の精神を十分に学べる施設となっている。

Wakayama Prefectural Animal Welfare Center

Location	Kaiso-gun, Wakayama Prefecture
Completion	March 2000
Structure	Steel, RC/1 floor below-grade, 2above-ground floors
Site area	31,750m^2
Building area	4,255m^2
Total floor area	4,057.32m^2

This facility was planned as a center for development and dissemination of the knowledge necessary for contact with animals and for their husbandry. Overall adoption of individual dog dens for housed animals is on a first trial nationwide, and this is a place where the spirit of animal lovers can be learned.

正面外观
正面外観
External view on the front

展庭的二层天桥
展示室の 2 階ブリッジ
Bridge on the 2nd floor in exhibition room

室外游具"动物塔"
屋外遊具「アニマルタワー」
Outdoor play structure 'Animal tower'

展厅
展示室
Exhibition room

玄海能源公园

所 在 地	佐贺县东松浦郡
竣工时间	2000 年 3 月
建筑结构	钢筋混凝土结构，部分为钢结构 / 地上 6 层，屋顶间 1 层
用地面积	887954.70m²
占地面积	5404.82m²
建筑面积	9802.53m² (只做基本设计)

本建筑作为九州电力的展示馆位于能够眺望到玄界滩的地理位置。该展示馆是由两部分所构成，它们分别是介绍电力及原子能发电的科学馆“火之翼”和介绍九州地方文化的九州故乡馆“歌之翼”。

玄海エネルギーパーク

所 在 地	佐賀県東松浦郡
竣 工 年	2000 年 3 月
規　　模	RC 造、S 造／地上 6 階、塔屋 1 階
敷地面積	887,954.70m²
建築面積	5,404.82m²
延床面積	9,802.53m²(基本設計のみ)

九州電力の展示館「玄海エネルギーパーク」は玄界灘を望む位置にある。展示館は電力と原子力発電を紹介するサイエンス館「火のウィング」と九州の地域文化を紹介する九州ふるさと館「うたのウィング」の 2 つのブロックで構成される。

Genkai Energy Park

Location	Higashi Matsuura-gun, Saga Prefecture
Completion	March 2000
Structure	RC, Steel/6 above-ground floors
Site area	887,954.70m²
Building area	5,404.82m²
Total floor area	9,802.53m²(Only Basic Design)

Kyushu Power Company's "Genkai Energy Park" Exhibition Building is located at a spot for viewing the sea at Genkai Nada. The exhibition building is composed of two blocks, the science building, called "Fire Wing", that introduces electrical and nuclear power generation, and the hometown Kyushu building, "Song Wing", that introduces the regional culture of Kyushu.

正面是科学馆"火之翼"
正面はサイエンス館「火のウイング」
The Science Center "Wing of Fire" on the front

入口
エントランス
Entrance area

观览室
展望ルーム
Observatory room

常陆海滨公园“卵之森”

所 在 地	茨城县常陆中市
竣工时间	1993 年
建筑结构	木结构，部分为钢结构/一周长80m

该设施位于茨城县东部的日立海滨公园的松林之中。以由家长带领着的小孩为主要使用对象。卵的造型物布置在设施中，力图营造出充满幻想的空间氛围。

ひたち海浜公園「たまごの森」

所 在 地	茨城県ひたちなか市
竣工年月	1993 年
規　　模	木造・一部 S 造／一周 80m

茨城県東部、ひたち海浜公園に設置された幼児のためのエリア、たまごをテーマにさまざまなたまごの物語をイメージした造形物が森の中に設置された。

Hitachi Seaside Park ‘Woodland of Eggs’

Location	Hitachinaka City, Ibaraki Prefecture
Completion	1993
Structure	W, part Steel/80m

A structure with an "egg" theme from various stories of eggs was built in the forest as a play area for infant children in this seaside park in eastern Ibaraki Prefecture.

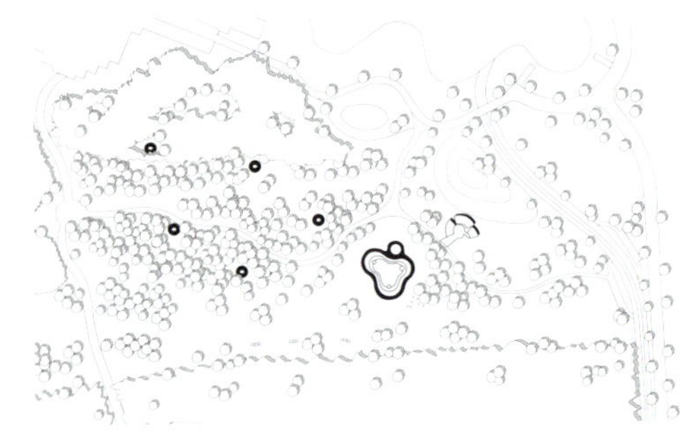

各种形状的"鸡蛋"分散点缀在森林中
さまざまな形のユニークな「たまご」が森の中に点在する
Various shaped interesting "eggs" are provided in the woodland

安浦地区沿道路绿化带“水之丘”

所 在 地	神奈川县横须贺市
竣工时间	1993 年
建筑结构	钢筋混凝土结构
用地面积	45000m²

本次在安浦新港修建的全新的以水为主题的公园设计是作为该项整备工程的中心而进行的。公园的水面积有690m²，采用了水从巨大的梁上落下的形式。落水长11m,宽3.3m。水面从草坪广场流向舞台，从而使落水显得更加壮观。

安浦地区沿道緑地「水の丘」

所 在 地	神奈川県横須賀市
竣工年月	1993 年
建築構造	RC 造
敷地面積	45,000m²

その中心的なスペースとして、安浦の新港に新しい水をテーマとした公園がつくられた。水面積は690㎡で、大きな梁から水が落下する形式のものである。落水の長さは11m、高さ3.3mである。水面は芝生広場からステージに上がっており、それがまた落水をより大きく見せている。

Yasuura Roadway Landscaping

Location	Yokosuka City, Kanagawa Prefecture
Completion	1993
Structure	RC
Site area	45,000m²

This park, on a water theme, was created as the central space for the new port in Yasuura. With a water area of 690sqm, waterfall length of 11m and a height of 3.3m, it has the shape of water falling from massive beams. The surface of the water ascends to the stage from the lawn square, and makes the waterfall even look larger.

摄影：环境设计研究所

撮影：環境デザイン研究所
Photo: Environment Design Institute

纸百科

所 在 地	东京都港区银座
竣工时间	2000 年 12 月
建筑结构	RC(现存建筑)，一层的室内室外重新装修
建筑面积	约 200m^2
获奖经历	2001 年展示设计奖

本建筑是专门为图像设计师等纸张使用者开设的纸张样品展示场。在空间设计上采用简洁的设计手法，力图衬托和展示出纸张所具有的独特魅力。

紙百科

所 在 地	東京都港区銀座
竣 工 年	2000 年 12 月
規　　模	RC 造（既存建物）／1 階内部及び 1 階外装部を改修
延床面積	約 200m^2（1 階改修部分のみ）
受 賞 歴	2001.ディスプレイデザイン賞

紙の卸会社のグラフィックデザイナー等のユーザーのための紙の見本展示場である。シンプルな空間構成の中にサンプルとしての紙の美しさを演出した空間構成としている。

Paper House

Location	Ginza, Minato-ku, Tokyo
Completion	December 2000
Structure	RC (existing building); interior and exterior of the first floor - refurbished
Total floor area	Approx. 200m^2 (refurbished area)
Awards	2001 Display Design Award

This is a paper sample exhibition site of a wholesale company of paper for such users as graphic designers. The beauty of paper sample is presented in simple spatial structure.

纸张在统一的白色调室内空间里,其颜色显得更加鲜亮
白で統一された内部に紙の色鮮やかさが映える
Vivid colors of papers look impressive in the white interior

如同白色箱体一样的外观
白い箱のような外観
Exterior appearance like a white box

入口外观
アプローチ外観
View of approach

浴室内部和百页顶棚
浴室内部とルーバー天井
Inside bath room and louvered ceiling

摄影：环境设计研究所

撮影：環境デザイン研究所
Photo: Environment Design Institute

弘法汤

所 在 地	石川县七尾市
竣工时间	2002 年 10 月
建筑结构	钢筋混凝土结构 / 地上 3 层
用地面积	820.6m²
占地面积	431.2m²
建筑面积	688.15m²
获奖经历	2003 年石川县建筑奖

本建筑是位于能登半岛入口处的七尾市的公共浴场。其1层部分是浴场，2层部分是浴场老板家的自宅。浴场是由更衣室和浴室以及夹在它们之间的中庭所构成。浴场空间因中庭的存在而显得舒适开敞，从而使来这里洗浴的人们可以得到充分的放松和休息。

弘法湯

所 在 地	石川県七尾市
竣 工 年	2002 年 10 月
規 模	RC 造／地上 3 階
敷地面積	820.6m²
建築面積	431.2m²
延床面積	688.15m²
受 賞 歴	2003.石川県建築賞

能登半島の入口に位置する七尾市の大衆浴場である。1階部分は浴場、2階部分は経営者の住居である。浴場は更衣室部分と浴室を中庭をはさんで対置させ、中庭の美しさによってゆったりとした心身の開放感を促す空間演出が意図された。

Bath House “Kobo-Yu”

Location	Nanao City, Ishikawa Prefecture
Completion	October 2002
Structure	RC/3 above-ground floors
Site area	820.6m²
Building area	431.2m²
Total floor area	688.15m²
Awards	2003 Ishikawa Architecture Award

A bath house in Nanao City, located at the entrance of Noto Peninsula. On the first floor are the baths and on the second floor is the managers’ residence. As for the baths, the spatial staging promotes relaxed and liberating bodily and mental sensations though the beauty of an inner court, with the changing room portion and baths counterpoised around this court.

御代田学校

所 在 地	长野县北佐久郡
竣工时间	2002 年 7 月
建筑结构	钢筋混凝土结构，部分为钢结构 / 地上 2 层
用地面积	12000m²
占地面积	3945m²
建筑面积	4781m²

该建筑是为住民设置的终生学习机构，是包括能容纳 500 人的大厅、图书馆和绳文博物馆的综合设施。为了与城市的街道景观和浅间山的自然景观相协调，建筑采用了当地传统的坡屋顶的建筑形式。

エコールみよた

所 在 地	長野県北佐久郡
竣 工 年	2002 年 7 月
規 模	RC 造一部 S 造／地上 2 階
敷地面積	12,000m²
建築面積	3,945m²
延床面積	4,781m²

町民のための生涯学習施設であり、500 人収容のホール、図書館、縄文博物館の複合施設である。街並みと浅間山の景観を考慮して、切妻屋根とした。地域継承型の建築である。

Ecole Miyota

Location	Kitasaku-gun, Nagano Prefecture
Completion	July 2002
Structure	RC part Steel/2 above-ground floors
Site area	12,000m²
Building area	3,945m²
Total floor area	4,781m²

This is lifetime study facility for town residents, a multi-complex with a 500 person capacity hall, library and Jomon Museum. With its gabled roof it contemplates the streets and view of Mt. Asama. This is construction of the local vernacular type.

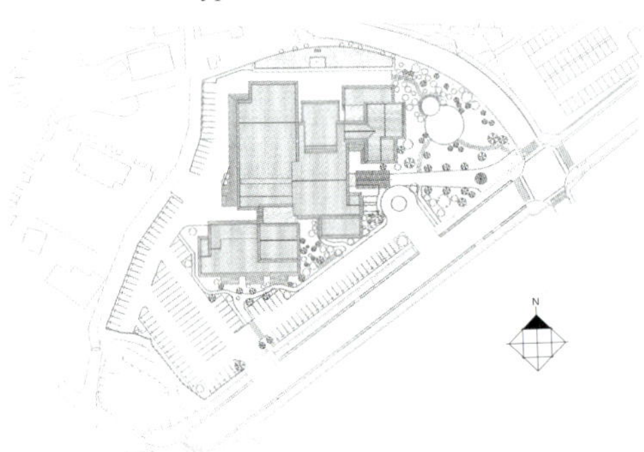

主入口
メインエントランス
Main entrance

入口处的中庭
エントランスアトリウム
Entrance atrium

摄影：相原功
撮影：相原功
Photo: Isao Aihara

福井市文化设施项目

所 在 地　福井县福井市
竣工时间　2004年3月
建筑结构　钢结构，部分为钢框架钢筋混凝土结构/地上9层，地下1层
用地面积　447.84m²
占地面积　382.08m²
建筑面积　2478.49m²

该建筑是以促进和搞活地方文化为目的而建造的文化设施，位于福井市中心区的证券公司大厦的用地范围内。以能容纳100人的小剧场为中心的该文化建筑，希望借此能够推动市民的艺术文化活动。

福井まちなか文化施設プロジェクト

所 在 地　福井県福井市
竣 工 年　2004年3月
規 模　S造一部SRC造／地上9階地下1階
敷地面積　447.84m²
建築面積　382.08m²
延床面積　2,478.49m²

福井市の中心街の証券会社ビル用地に地域活性化をはかる拠点としてつくられる文化施設プロジェクトである。収容100名の小劇場を中心に市民の芸術文化活動の中心となることが期待されている。

Fukui Cultural Complex Project

Location　Fukui City, Fukui Prefecture
Completion　March 2004
Structure　Steel, part SRC /1floor below-grade, 9 above ground floors
Site area　447.84m²
Building area　382.08m²
Total floor area　2,478.49m²

A cultural facility building project on the site of a securities firm in the central district of Fukui City that was created as a foothold for regional area activation. Centered on a 100 person capacity mini theater, it is expected to be a center of art and cultural activities for residents.

上海闵行区莘庄北广场综合办公楼项目

所 在 地　中国上海市闵行区
竣工时间　2005 年
建筑结构　钢筋混凝土结构，部分为钢结构/地上24层，地下1层（税务局·综合政府机构楼），地上14层，地下1层（海关·商检楼），地上17层，地下1层（商务写字楼），地上1层，地下1层（展示场）
用地面积　42300m²
建筑面积　11918m²
建筑面积　116815m²
共同设计　现代设计集团、华东设计院

本建筑在设计上借鉴了中国上海的三合院的住宅形式，即由24层税务局·综合政府机构楼、14层海关·商检楼和17层商务写字楼等三座门形的办公楼围合而成，并且通过连廊将它们连接起来，并在由三座办公楼围合而成的市民广场的地下设置了展示空间。

上海閔行区オフィスプロジェクト

所 在 地　中国上海市閔行区
竣 工 年　2005 年
規　　模　RC造一部S造／地上24階地下1階（税務局・合同庁舎棟）、地上14階地下1階（税関・商検棟）、地上17階地下1階（商務オフィス棟）、地上1階地下1階（展示場）
敷地面積　42,300m²
建築面積　11,918m²
延床面積　116,815m²
共同設計　現代設計集団 華東設計院

上海市閔行区オフィスセンタービルは中国上海の3合院という住宅形式を引用し、24階建ての税務局合同庁舎棟、14階建ての税関商検棟、17階建ての商務棟の3つの門状のオフィス棟で構成されているそれぞれの棟はブリッジで連結され、3棟に囲まれた市民広場の下には展示場が設けられている。

Shanghai Minghang Xinzhuang Office Building and Community Plaza Project

Location　Min-hang District, Shanghai, China
Completion　2005
Structure　RC part Steel/1 floor below-grade, 24 above-ground floors (Customs and Commerce Building); 1 floor below-grade, 14 above-ground floors (Tax Office Building); 1 floor below-grade, 17 above-ground floors (Commercial Building); 1 floor below-grade, 1 above-ground floor (Exhibition Area)
Site area　42,300m²
Building area　11,918m²
Total floor area　116,815m²
Co-operative Design　Shanghai Xian Dai Architectural Design Group East China Architectural Design and Research Institute

Xinzhuang Office Center Building in Shanghai, China, employs Shanghai's Sangoin residential form and is composed of three gate-like office blocks of 24 stories: a tax office joint government building block, a 14 story custom and commercial inspection block and a 17 story commercial business block, with each section connected by bridge. An exhibition site is established under the public square that is surrounded by the three blocks.

仙田满
Mitsuru Senda

作者简介

主要经历

1941	横滨市出生
1964	东京工业大学建筑学科毕业
1964-1968	菊竹清训建筑设计事务所
1968	环境设计研究所创立 所长
1982	工学博士（东京工业大学）
1984-1987	琉球大学教授
1988-1992	名古屋工业大学教授
1989-2002	宋庆龄基金会理事
1992-	东京工业大学教授
1999-2001	日本建筑学会副会长
2001-2003	日本建筑学会会长

主要获奖

1978	每日设计奖(业绩)
1987	国际交通安全学会奖（《儿童的游戏环境》）
1988	B.C.S 奖(滨松科学馆)
1993	日本建筑学会霞关高层建筑纪念奖(《关于都市空间中的儿童的游戏环境开发的研究》)
1996	日本造园学会奖(茨城县自然博物馆) 商业环境设计奖优秀奖(藤野艺术之家)
1997	日本建筑学会奖(爱知县儿童综合中心) IOC.IAKS 奖银奖（东京辰巳国际游泳馆） 东京建筑奖优秀奖（茨城县自然博物馆优秀奖）
2000	都市景观大奖(富岩运河环水公园)
2001	IOC.IAKS 奖金奖(但马球场)
2003	国际建筑 IAA 奖金奖(但马球场)

主要著作

1984	《儿童的游戏环境》(筑摩书房)
1985	《游戏环境的设计》(鹿岛出版会)
1986	《现代建筑空间和方法》(同朋舍出版)
1988	《儿童和住居》(住居的图书馆)
1990	《Design of Children's Play Environment》(McGraw Hill)
1992	《儿童和游戏》(岩波新书)
1998	《环境设计的方法》(彰国社)游戏构造（柏书房） 《为了儿童的游戏空间》(市谷出版社）
2001	《为了幼儿的环境设计》(世界文化社)
2002	《环境设计的展开》(鹿岛出版会)

中国的业绩

1985	宋庆龄基金会成员理事
1990	宋庆龄基金会理事
1993	北京科普乐园(中国北京宋庆龄基金会)设计完成
1995	北京科普乐园第二期设计完成
2001	同济大学讲演《以地球环境建筑的设计为己任》
2002	北京奥林匹克公园国际设计邀请赛参赛
2003.3	上海建筑学会邀请讲演会 邀请讲演
2003.4	同济大学建筑与城市规划学院客员教授
2003.6	上海旗忠森林体育城网球中心国际设计邀请赛 一等奖中标
2003.7	上海闵行区莘庄北广场综合办公楼国际设计邀请赛 一等奖中标
2003.10	广东省佛山市体育馆国际设计邀请赛 一等奖

株式会社 环境设计研究所
邮政编码:106-0032 日本国东京都港区六本木 5-12-22
电话:0081-3-5575-7171
传真:0081-3-5562-9928
上海市延安西路 688 号楼 3-14D 室 688#
电话:021-62261919
传真:021-62255566
E-mail:msenda@ms-edi.co.jp
Website:http://www.ms-edi.co.jp

著者プロフィール

経歴

1941	横浜市に生まれる
1964	東京工業大学建築学科卒
1964-1968	菊竹清訓建築設計事務所
1968	環境デザイン研究所創設 所長
1982	工学博士（東京工業大学）
1984-1987	琉球大学教授
1988-1992	名古屋工業大学教授
1989-2002	宋慶齢基金会理事
1992-	東京工業大学教授
1999-2001	日本建築学会副会長
2001-2003	日本建築学会会長

受賞

1978	毎日デザイン賞（業績）
1987	国際交通安全学会賞（「こどものあそび環境」）
1988	B.C.S 賞（浜松科学館）
1993	日本建築学会霞ヶ関ビル記念賞（「都市空間におけるこどものあそび環境開発に関する研究」）
1995	ディスプレイデザイン賞（茨城県自然博物館）
1996	日本造園学会賞（茨城県自然博物館） 商環境デザイン賞優秀賞（藤野芸術の家）
1997	日本建築学会賞（愛知県児童総合センター） IOC.IAKS賞銀賞（東京辰巳国際水泳場） 東京建築賞優秀賞（茨城県自然博物館）
2000	都市景観大賞（富岩運河環水公園）
2001	IOC.IAKS 賞金賞（但馬ドーム）
2003	国際建築アカデミー・グランプリ受賞（但馬ドーム）

著作

1984	こどものあそび環境（筑摩書房）
1985	あそび環境のデザイン（鹿島出版会）
1986	現代建築　空間と方法　仙田满（同朋舎出版）
1988	こどもと住まい（住まいの図書館出版局）
1990	Design of Children's Play Environment（McGraw Hill）
1992	こどもとあそび（岩波新書）
1998	環境デザインの方法（彰国社） プレイストラクチャー（柏書房） こどものためのあそび空間（市ヶ谷出版社）
2001	幼児のための環境デザイン（世界文化社）
2002	環境デザインの展開（鹿島出版会）

中国での実績

1985	宋慶齢基金会評議員
1990	宋慶齢基金会理事
1993	北京科普楽園第 1 期完成（北京宋慶齢基金会）
1995	北京科普楽園第 2 期完成
2001	同済大学（上海）講演「地球環境建築の展望」
2002	北京オリンピックグリーン国際デザインコンペ参加
2003.3	上海建築業界交流講演会（テーマ：日本建築界と万国博覧会）
2003.4	同済大学建築・都市計画学科客員教授として講演「万国博覧会に向けて」
2003.6	上海テニスセンター国際デザインコンペ最優秀（2005 年竣工予定）
2003.7	上海オフィスビル国際デザインコンペ最優秀（2005 年竣工予定）
2003.10	広東省佛山市体育館国際デザインコンペ一等賞

株式会社 環境デザイン研究所
住所：106-0032 東京都港区六本木 5-12-22
TEL：03-5575-7171
FAX：03-5562-9928
E-mail: msenda@ms-edi.co.jp
Website: http://www.ms-edi.co.jp

Profile

Personal History

1941 Born in Yokohama, Kanagawa Pref., Japan
1964 Graduated from the Department of Architecture, Tokyo Institute of Technology
1964-1968 Worked for Kiyonori Kikutake Architects and Associates
1968 Established Environment Design Institute as the president
1982 Took a doctorate in engineering at Tokyo Institute of Technology
1984-1987 Professor at Ryukyu University
1988-1992 Professor at Nagoya Institute of Technology
1989-2002 Committee member for Song-Qing-Ling Foundation
1992- Professor at Tokyo Institute of Technology
1999-2001 Vice president of Architectural Institute of Japan
2001-2003 President of Architectural Institute of Japan

Prizes

1978 Mainichi Design Award for works of children's play environment
1987 International Association of Traffic and Safety Sciences Award for his book, "Children's Play Environment"
1988 BCS Award for Hamamatsu Science Museum
1993 The Kasumigaseki Building Memorial Prize of Architectural Institute of Japan for his 'Research of Development of Children's Play Environment in Urban Area'
1995 Display Design Award for Ibaraki Nature Museum
1996 Japanese Institute of Landscape Architecture Prize for Ibaraki Nature MuseumJCD Award for Fujino Workshop for Art
1997 Architectural Institute of Japan Prize for Aichi Children's CenterIOC/IAKS Award Silver Prize for Tokyo Tatsumi International Swimming PoolTokyo Architecture Award Superior Award for Ibaraki Nature Museum
2000 Urban Landscape Award for Fugan Canal Park
2001 IOC/IAKS Award Gold Prize for Tajima Dome
2003 IAA Award Grand Prix for Tajima Dome

Publications

1984 "Children's Play Environment", Chikuma Shobo.
1985 "Design of Children's Play Environment", Kajima Shuppan.
1988 "Children and their living", SUMAI Library Publishing Company.
1990 "Design of Children's Play Environment", McGraw-Hill.
1992 "Children and their play", Iwanami Shobo.
1998 "Methods of Environment Design", Shokokusha.
2001 "Environment Design for Infants", Sekai-Bunkasha.
2002 "Development of Environment Design", Kajima Shuppan.

Achievements in China

1985 Council member of Song-Qing-Ling Foundation
1990 Committee member of Song-Qing-Ling Foundation
1993 Completion of the 1st phase design of Beijing Song-Qing-Ling Children's Science Park
1995 Completion of the 2nd phase design of Beijing Song-Qing-Ling Children's Science Park
2001 Lecture in Tongji University, Shanghai–"Towards Earth Environmental Architecture"
2002 Participated in the international design competition for Beijing Olympic Green
2003.3 Symposium and lecture in Shanghai (lecture about Japanese architecture industry and the world expo)
2003.4 Visiting professor at Department of Architecture and Urban Planning of Tongji University and gave a lecture titled, "The Way to the World Expo"
2003.6 Received the best prize for the international design competition for Shanghai Qizhong Forest Sports City Tennis Center (planned to be completed in 2005)
2003.7 Received the best prize for the international design competition for Shanghai Xinzhuang Office Building and Community Plaza (planned to be completed in 2005)
2003.10 Received the first prize for the international design competition for Foshan City Gymnasium

Environment Design Institute
Add: 5-12-22, Roppongi, Minato-ku, Tokyo 106-0032, JAPAN
Tel: +81-3-5575-7171
Fax: +81-3-5562-9928
E-mail: msenda@ms-edi.co.jp
Website: http://www.ms-edi.co.jp